Praises for The Exponential Era

"*The Exponential Era* turns strategic planning from a stagnant limited application exercise to an active thoughtful process that can yield benefits for all companies and executives. Every company leader can find a gem in *The Exponential Era* to apply to their business big or small."

Michael Splinter, Chairman of the Board, NASDAQ and Retired Chairman and Chief Executive Officer, Applied Materials

"I count this among the very best business books I have read. The authors have managed to synthesize a vast array of thinking and methodologies and deployed them in a practical and easily understood planning process (SPX) that addresses today's exponential pace of change ... and they have thrown in a bonus chapter on disruptions we should all be prepared for in the future.

Strategic Planning for the Exponential Era (SPX) holds great promise to address the extreme challenge of identifying future opportunities and ensuring that all ingredients are in place to launch the products and services that customers will love at just the right future moment.

Highly recommended for today's business leaders. Espindola and Wright have delivered an extremely valuable guide to the future of business planning ... and that future is now!"

James B. Stake, former Executive Vice President, Enterprise Services, 3M Company, Chairman, Ativa Medical Corporation and Member of Board of Directors, public and private companies

"*The Exponential Era* is a must-read for business leaders, entrepreneurs, and virtually anyone navigating our highly complex and rapidly changing world. David Espindola and Michael W. Wright offer a great model for a thoughtful and logical way to develop a strategy for environments dominated by innovation, disruption, and rapidly changing market forces. Along the way – Espindola and Wright draw on the already broad subject area to give the reader an idea of not only where we have been but where the art of strategy formation and decision-making is going. Entertaining, practical, well-researched, I strongly recommend this book, a primer for navigating the future."

General (Ret. 4 Star) Joseph L. Votel, President and CEO, Business Executives for National Security (BENS)

"A century ago, a convergence across domains ushered in unprecedented advancements in human development. A look at history is very instructive, as several dynamics from that period have the potential to emerge once again – the biggest being the opportunity for convergence. In this context, convergence refers to a virtuous cycle where events in one domain spur action in another. As a futurist, focusing on the intersections provides a window into the future. Visualizing the building blocks and how they converge is the most effective way to navigate a very uncertain future. *The Exponential Era* does a great job of not only describing exponential technologies, but how they likely converge to transform our world. Education and dialog are critical to creating a constructive future – this book contributes to that end."

Frank Diana, Managing Partner, Futurist, TATA Consultancy Services

---

"*The Exponential Era* is an essential read for our times. Our values, curiosity and a vision that champions change and improvement (Always Get Better!) embraces many of the concepts captured in the SPX methodology outlined in the book. By monitoring horizons, initiating multiple small experiments and scaling rapidly, we were able to achieve a higher level of profitability and stability despite the chaotic changes all around us. Dynamic planning processes, like those described in this book, when coupled with great cultures, work!"

John Puckett, Owner of Punch Pizza and Co-founder of Caribou Coffee

---

"*The Exponential Era* will help corporations and its leaders see the technical tornados coming and have the courage to experiment and walk toward the storms, emerging on a new path to a successful exciting future. The Exponential Era is a scary place - learn to embrace the change, enjoy it, and thrive."

Ronald Peterson, PhD, former Vice President of Technology, Honeywell Corporation and Author, *Gardeners of the Universe*

"The new conventional wisdom highlights the increasing rate of change of business processes and the disruption of product markets driven by Moore's law. The challenge that many companies face is that by the time competitive threats driven by Moore's law become visible, it is too late for them to respond. *The Exponential Era* provides a roadmap for strategic decision making in a digital world. It is a must read for top management teams who recognize the importance of disrupting themselves before they are disrupted."

Stefanie Ann Lenway, Dean, Opus College of Business, University of St. Thomas

"Whether your business is a 'flash boiled frog,' a 'fast and furious gazelle,' a 'disruptive unicorn,' or a 'dancing elephant,' you will find a lot of useful terms and insightful ideas in this book about the exponential growth of digital platforms and technologies that have overturned established companies and entire industries. The book offers a new approach to strategic planning in this exponential era – SPX – that is fast, flexible and iterative, using design thinking and OODA (Observe-Orient-Decide-Act) methods to help companies become the disrupters rather than the disrupted. Read this book and let the experimentation begin!"

Thomas Fisher, Professor, Director of the Minnesota Design Center and Dayton Hudson Chair in Urban Design, College of Design at the University of Minnesota.

"Many organizations cling to what eventually kills them. This book spotlights the root causes and prescribes a new operating system calibrated for thriving in the exponential era. It's likely to be the book you wish you'd read five years ago!"

Elvin Turner, Author, *Be Less Zombie: How Great Companies Create Dynamic Innovation, Fearless Leadership and Passionate People*

"Michael Wright and David Espindola have written a Future Shock for the 2020s and beyond. This is a guide not simply for the engineering community, but for the broader society of all organizations, non-profits, and individuals. They do more than explain new technologies and systems. They explain how these systems are interrelated and will affect our lives. In the end, SPX is a process for asking the right questions."

Todd Lefko, President, International Business Development Council and Chairman, Russian-American Business and Culture Council (RABCC)

"If you want to know how culture can add to exponential value, look no further. This is the guide and much more."

Marcus Kirsch, Author, *The Wicked Company*

"The authors have been consistently right about capitalizing on 'What's next?' The COVID-19 environment has challenged CEOs and entrepreneurs to immediately rethink and accelerate their Strategy-2-Execution and Ideation-2-Monetization value chains. *The Exponential Era* is an optimistic must read and SPX is a practitioners' playbook to become an agile, innovative, sustainable and profitable enterprise."

Murad S. Velani, CEO of Bluespire

THE EXPONENTIAL ERA

A Note from the Series Editor

Welcome to the brand-new Wiley-IEEE Press Series on Technology Management, Innovation, and Leadership!

The IEEE Press imprint of John Wiley & Sons is well known for its books on technical and engineering topics. This new series extends the reach of the imprint, from engineering and scientific developments to innovation and business models, policy and regulation, and ultimately to societal impact. For those who are seeking to make a positive difference for themselves, their organization, and the world, technology management, innovation, and leadership are essential skills to hone.

The world today is increasingly technological in many ways. Yet, while scientific and technical breakthroughs remain important, it's connecting the dots from invention to innovation to the betterment of humanity and our ecosphere that has become increasingly critical. Whether it's climate change or water management or space exploration or global healthcare, a technological breakthrough is just the first step. Further requirements can include prototyping and validation, system or ecosystem integration, intellectual property protection, supply/value chain set-up, manufacturing capacity, regulatory and certification compliance, market studies, distribution channels, cost estimation and revenue projection, environmental sustainability assessment, and more. The time, effort, and funding required for realizing real-world impact dwarfs what was expended on the invention. There are no generic answers to the big-picture questions either; the considerations vary by industry sector, technology area, geography, and other factors.

Volumes in the series will address related topics both in general—e.g., frameworks that can be applied across many industry sectors—and in the context of one or more application domains. Examples of the latter include transportation and energy, smart cities and infrastructure, and biomedicine and healthcare. The series scope also covers the role of government and policy, particularly in an international technological context.

With 30 years of corporate experience behind me and about five years now in the role of leading a Management of Technology program at a university, I see a broad-based need for this series that extends across industry, academia, government, and nongovernmental organizations. We expect to produce titles that are relevant for researchers, practitioners, educators, and others.

I am honored to be leading this important and timely publication venture.

Tariq Samad
Senior Fellow and Honeywell/W.R. Sweatt Chair in Technology Management
Director of Graduate Studies, M.S. Management of Technology
Technological Leadership Institute | University of Minnesota
samad@ieee.org

IEEE Press Series on Technology Management, Innovation, and Leadership

THE EXPONENTIAL ERA

STRATEGIES TO STAY AHEAD OF THE CURVE IN AN ERA OF CHAOTIC CHANGES AND DISRUPTIVE FORCES

David Espindola

Michael W. Wright

IEEE**TEMS**
Technology & Engineering
Management Society

IEEE PRESS

WILEY

The right of David Espindola and Michael W. Wright to be identified as the authors of this work has been asserted in accordance with law.

Registered Office
John Wiley & Sons, Inc., 111 River Street, Hoboken, NJ 07030, USA

Editorial Office
111 River Street, Hoboken, NJ 07030, USA

For details of our global editorial offices, customer services, and more information about Wiley products visit us at www.wiley.com.

Wiley also publishes its books in a variety of electronic formats and by print-on-demand. Some content that appears in standard print versions of this book may not be available in other formats.

Limit of Liability/Disclaimer of Warranty
In view of ongoing research, equipment modifications, changes in governmental regulations, and the constant flow of information relating to the use of experimental reagents, equipment, and devices, the reader is urged to review and evaluate the information provided in the package insert or instructions for each chemical, piece of equipment, reagent, or device for, among other things, any changes in the instructions or indication of usage and for added warnings and precautions. While the publisher and authors have used their best efforts in preparing this work, they make no representations or warranties with respect to the accuracy or completeness of the contents of this work and specifically disclaim all warranties, including without limitation any implied warranties of merchantability or fitness for a particular purpose. No warranty may be created or extended by sales representatives, written sales materials or promotional statements for this work. The fact that an organization, website, or product is referred to in this work as a citation and/or potential source of further information does not mean that the publisher and authors endorse the information or services the organization, website, or product may provide or recommendations it may make. This work is sold with the understanding that the publisher is not engaged in rendering professional services. The advice and strategies contained herein may not be suitable for your situation. You should consult with a specialist where appropriate. Further, readers should be aware that websites listed in this work may have changed or disappeared between when this work was written and when it is read. Neither the publisher nor authors shall be liable for any loss of profit or any other commercial damages, including but not limited to special, incidental, consequential, or other damages.

Library of Congress Cataloging-in-Publication Data:
Names: Espindola, David, author. | Wright, Michael W., author.
Title: The exponential era: strategies to stay ahead of the curve in an era of chaotic
  changes and disruptive forces / David Espindola, Michael W. Wright.
Description: Hoboken, New Jersey : Wiley-IEEE Press, [2021] | Includes
  index.
Identifiers: LCCN 2020044905 (print) | LCCN 2020044906 (ebook) | ISBN
  9781119814047 (hardback) | ISBN 9781119746522 (adobe pdf) | ISBN
  9781119746539 (epub)
Subjects: LCSH: Business. | Business enterprises–Technological
  innovations.
Classification: LCC HF5351 .E76 2021 (print) | LCC HF5351 (ebook) | DDC
  338/.064–dc23
LC record available at https://lccn.loc.gov/2020044905
LC ebook record available at https://lccn.loc.gov/2020044906

Cover Design: Wiley
Cover Image: © ChrisHepburn/Getty Images

Set in 10/12pt JansonText by SPi Global, Pondicherry, India

SKY10024192_012621

First and foremost, the glory belongs to the Lord.
Second, I dedicate the book to the people that I love:
Dawn, Kayla, Andre, Nair, Lidia, Juliana and Clarice.

–David

To all of my daughters who are, and will be,
leaders in our global quest to improve our only home, earth.
And to my son for being a net contributor to a better future
in business and life.

–Michael

CONTENTS

ABOUT THE AUTHORS

David Espindola is a Partner of Intercepting Horizons, a strategic advisory services firm. He also serves on the advisory board of the Technological Leadership Institute at the University of Minnesota.

As a former CIO, he has developed and implemented global technology strategies to drive business growth and has been part of the leadership team that delivered a business transformation program resulting in 5× revenue growth to more than $4 billion. Previously, as a consulting leader, he managed large and complex engagements at strategic accounts and sold several million dollars in software and services.

Before that, Espindola held leadership roles at two fast-growing tech startups in Silicon Valley. One spun off into a successful business still operating today. The other grew 5× during his tenure, reaching over $400 million in revenues, resulting in a successful IPO.

Espindola was awarded an MBA from the American University and a BS in Engineering Management from the University of North Dakota. He is also a graduate of Stanford University's Executive Education program.

Michael W. Wright is a Partner of Intercepting Horizons and the author of the acclaimed *New Business Normal*. He is an active board member, a Sr. Fellow at the University of Minnesota (UMN), Chairman Emeritus of the Advisory Board for the UMN Technological Leadership Institute, and formerly an adjunct professor at Carlson School of Management where he taught strategic leadership.

In his extensive career, he has served as CEO, COO, and CMO, board member, author, business professor, futurist, speaker, and proven entrepreneur. He has delivered more than three decades of global P&L leadership for diverse public and privately held early stage to mature technology enterprises at scale. Throughout his career he has focused on leading-edge semiconductor equipment and materials, fluid flow and filtration, instrumentation, software, cost modeling, and simulation product portfolios.

Wright is a graduate of the U.S. Navy Nuclear Power Program, and a graduate of the Rochester Institute of Technology and Kellogg Executive Programs.

ACKNOWLEDGMENTS

This book would not have been possible without the wisdom, support, and guidance of many incredible people. First, the authors would like to express our gratitude to our families for their love and patience as we spent endless hours immersed in our thoughts in front of a computer. We would also like to thank all the people at Wiley-IEEE Press who worked diligently to make this project successful. Our special thanks to Dr. Tariq Samad for his guidance, reviews, and insights in the early stages of this venture, and to Mary Hatcher for helping us through the publishing process.

We would like to thank all the reviewers who kindly provided thoughtful input that further guided the development of the book, including Patrick McKinney, Don O'Shea, Jeff Hand, Dr. Alfred Marcus, Dr. Juan Bardina, and Dr. Rob Bodor.

Our special thanks to the University of St. Thomas, and in particular to Stefanie Lenway, Lisa Abendroth, and Carleen Kerttula for believing in us and giving us an opportunity to share our thoughts on converging technologies shaping the Exponential Era, resulting in the conception of the Contex Conference. Our thanks also go to the Technological Leadership Institute (TLI) at the University of Minnesota, its leadership, staff, and the advisory board for the opportunity to turn our passion in technology and education into contributions, as small as they may be, that hopefully will have a positive impact in the education of future technology leaders.

We would like to acknowledge all the organizations and individuals who we have had the privilege to work and interact with throughout our careers. You have allowed us to learn, to experiment, to gather, and to synthesize a lifetime of knowledge and experience, much of which has been shared in the pages of this book.

Finally, we would like to thank all of our colleagues, partners, contributors, and customers at Intercepting Horizons who have supported us and who continue to inspire us to do our work.

FOREWORD

As the Commander of United States Central Command from 2016 until 2019, we confronted unprecedented complexity and change daily. If it was not the adversaries we faced in places like Iraq, Syria, and Afghanistan, it was the deep underlying tensions coursing through the region: corruption, disenfranchisement, poor governance, extreme poverty, toxic sectarian narratives, and malign influences. On a day-to-day basis, we navigated near-term problems reasonably well. Still, the region's complexity and sheer chaos challenged us in charting a long-term strategy that would preserve our national interests. Any achievements that we could obtain came on the backs of skilled military planners and policymakers using the best analytic and planning tools available. I suspect if you spoke with any of them, they would tell you that an appreciation and comprehensive understanding was crucial for prevailing in this region.

Complexity reigns in today's strategic environment. We see it in our domestic and foreign policy discussions. A hyper-enabled information environment that reshapes facts to new truths fogs clarity for business executives and government leaders. Add to this the emergence of artificial intelligence and machine learning – technologies that are as promising as they are terrifying. Today, the United States is locked in a global competition with a rising China and, to a lesser extent, with a resurgent Russia. The battle is not just about military or economic dominance. The contest is also about the influence and domination of emerging technologies, and the one who gets there first will likely write the rules and enforce the norms. The stakes could not be higher.

The Exponential Era – Strategies to Stay Ahead of the Curve in an Era of Chaotic Change and Disruptive Forces by Michael Wright and David Espindola is a groundbreaking contribution to sense-making and strategy development. With a revisit to various planning tools and approaches employed over the last several decades, this volume offers a fresh new approach to understanding and assessing complexity and getting ahead of the curve.

Importantly, this book offers a new methodology known as "SPX" – short for Strategic Planning for the Exponential Era. SPX is a norm-busting approach focused on driving innovation and mapping risk, opportunities, and capabilities to create plans designed to stay ahead of exponential change. SPX is a comprehensive method of looking at exponential change caused by technology, operating environments, and conflicting interests.

The authors are highly experienced and well-qualified in this field. Michael Wright is a global senior high technology executive and strategist at scale. David Espindola is an accomplished technology executive and consultant. Together they are the founders of Intercepting Horizons, a strategic advisory service that focuses on teaching leaders and organizations to apply SPX to stay ahead of the change and complexity curve. Both serve on the Advisory Board for the University of Minnesota Technological Leadership Institute. I work with Michael on the Board of a small private company in Saint Paul, and through him, I have come to know David. Having spent a career dealing with complexity and chaos, I can attest that these two are onto something new, relevant, and exciting with SPX.

For those who may be concerned about reading a book on strategy written by technologists – rest easy. *The Exponential Era* is a logical and easy read, well-organized, and crisp in its overall presentation. I spent my military career around complexity and strategy development – I not only enjoyed reading this book, but I found it incredibly insightful and fresh in its overall approach. You will find it similarly satisfying.

Joseph L. Votel
General, US Army (Retired)

Introduction

I n the fall of 2018, we had the opportunity to engage in a discussion with a board member of a Fortune 500 logistics company. Having recently founded Intercepting Horizons, LLC we shared our thoughts on how we had entered a new era in which strategic trends and converging technology vectors were impacting organizations at unprecedented speeds. We called it the Exponential Era.

We discussed how existing planning cycles were no longer effective in this new environment, and how our services would focus on helping companies identify inflection points and develop an innovative strategic planning process that is responsive to fast-changing conditions. We would help companies pinpoint not only what strategic trends and converging technologies would impact their business, but equally important, when.

We felt confident in our ability to provide such services because we had learned over decades of successfully chasing Moore's law that it is possible to develop a time-enriched strategic planning process to intercept business inflection points and stay up with and sometimes ahead of exponential change.

This was enough to get the board member intrigued. He confided in us that his CEO had been contemplating these exact questions – when would these business inflection points impact their industry and business?

This informal conversation led to an introduction, and subsequently we started researching the company, its industry, the competitive forces, and the converging technologies that could represent a threat or an opportunity to this very large enterprise that for more than 100 years had been tremendously successful.

The Exponential Era: Strategies to Stay Ahead of the Curve in an Era of Chaotic Changes and Disruptive Forces, First Edition. David Espindola and Michael W. Wright.
© 2021 by The Institute of Electrical and Electronics Engineers, Inc.
Published 2021 by John Wiley & Sons, Inc.

What we discovered in researching new entrants to their industry and identifying technology vectors relevant to their business had us concerned.

The Threat of Being Amazoned and Uberized

Keep in mind that this company was the undisputed leader in its industry. They continued to experience year-over-year net revenue growth and healthy margins, rewarding investors with solid dividends, and repurchasing a significant amount of stock shares. The repurchase of shares might have been a sign of uncertainty about the company's prospects and an attempt to boost the share price in the short term. However, the leadership team displayed nothing but total confidence in the company's ability to stay competitive and grow.

What had us concerned, despite the company's position as a category king in its market, was the potential for disruption. For many years this company had operated under the following conditions:

1. Thousands of employees
2. Legacy culture, infrastructure, and profit pools
3. Relatively slow to move

The enormous size of this market and the potential for profits were very attractive to potential disruptive entrants with the following contrasting characteristics:

1. Access to cheap technology
2. A low-cost infrastructure that could be rented
3. Fast-moving

Indeed, what we found is that there were several exceptionally well-funded, fast-moving technology companies that were entering their markets.

It would be very disconcerting for any company to have its market leadership challenged by the likes of Amazon, Google, Uber, or other highly capitalized fast-moving companies. No one wants to be in that position.

The number of companies that have been "amazoned" is legendary.[1] An entire industry has been "uberized" due to Uber's unrelenting long-term investments in technology. Now, imagine being in the unenviable position of potentially being susceptible to competition from both Amazon and Uber at the same time – that is the threat level that this company was dealing with.

Our research had also revealed that the convergence of several emerging technologies could completely alter the market landscape that this company operated in. We leveraged Artificial Intelligence algorithms to mine patent databases and discovered, for example, that several large companies like Walmart and Amazon were investing in the ability to manufacture products on demand and on the move. The confluence of 5G, Internet of Things (IoT) and GPS capabilities, and the convergence of 3D printing with autonomous vehicles, is just one of many possible technology-driven disruptions on the horizon that could completely change their industry.

This additional background information helped us understand what might be driving the CEO's quest for knowing "when." Despite all their successes, this was a company in the eye of a storm, faced with disruptive forces and competitive pressures unlike any they had experienced before. It was vital for this company to get ahead of the inflection point. They needed to be nimbler and more aggressive than their competitors in targeting converging technology investments. They needed to overcome their history and get their entire culture ready for the disruption and transformation of their entire business.

The CEO was absolutely correct in his thinking as he sought to answer the question of "when." Bringing an innovative idea to market too early can be just as fatal as being too late. Think about Apple and the Newton personal assistant device.[2] Or Lady Florence Norman who introduced the motorized scooter circa 1916.[3] These were all great ideas, as proven by Apple's eventual success with the iPhone, and the motorized scooter craze we see around the world these days – they were just timed incorrectly.

Not too long after our discussions, the company announced that it was committing a significant amount of dollars in technology investments over the next five years.

This announcement was a positive sign. It showed that the company understood the challenges ahead. The fact that the CEO was seeking answers to the question of "when" told us he had a significant level of awareness that time enriched decision making is of the essence. However, as we will see in future chapters, we have witnessed and studied several companies that, notwithstanding full awareness and willingness to invest, were unable to make timely decisions, change their culture, and stay apace with the disrupting forces they faced.

The key point in sharing our interaction with this Fortune 500 company is to emphasize that any company, no matter how big or successful, can be subject to the disruptive forces of the Exponential Era. Newly formed ecosystems, powered by easy access to converging technologies, can provide the fuel to rapidly dislocate stagnant or poorly managed companies and industries. Our first word of caution is: do not be deceived by your success.

In the Exponential Era, it is very easy to be caught off guard, because what may look like calm seas, can quickly turn unexpectedly. If you can't foresee a major storm coming and can't prepare for it, it may be fatal.

Elephants Dancing to a Faster Tempo

As we will explore throughout the book, technology investments alone are not sufficient to achieve a successful adaptation to the speed and scale of change. The most challenging aspect of getting companies ready for the fast changes that characterize the Exponential Era is transforming the culture.

As Lou Gerstner said in *Who Says Elephants Can't Dance?* "Successful institutions almost always develop strong cultures that reinforce those elements that make the institution great. . . When that environment shifts, it is very hard for the culture to change. In fact, it becomes an enormous impediment to the institution's ability to adapt."[4]

Sometimes the only way to achieve a change in culture is by making a change at the CEO level, as was the case with the successful turn-around experienced by Microsoft. Before Satya Nadella, Microsoft's culture was often characterized as internally competitive and hostile. It prized showing that you were smart, even at the cost of creating hostility and preventing teamwork.

Microsoft was unable to keep up with the fast beat to which it was expected to dance. The company's prospects were quite dire for a while. It lost the mobile operating system war and almost failed to get traction in the cloud. It took bringing in a CEO that understood and emphasized culture to completely turn the company around. "There is something only a CEO uniquely can do, which is set that tone, which can then capture the soul of the collective. And its culture," said Nadella.[5]

Where Is My Crystal Ball?

"Predictions are hard, especially those that pertain to the future." This comical proverb was allegedly first expressed in Danish, but the author remains unknown.

Given the difficulty in making predictions, how do you deal with the uncertainties of the Exponential Era? How do you stay on top of all the emerging trends and converging technology factors? How you do spot inflection points in business before they happen? Rita McGrath has written an excellent book titled *Seeing Around Corners*, that expands on this subject. In the book, she contends that inflection points, though they may seem sudden, are not.[6] In an exponential curve, there is a long period of time in which the curve is basically flat, before the inflection point. Then it hits the elbow and goes straight up (see Figure I.1). It is during this flat period and the beginning of the elbow that, armed with the right perspectives and tools, smart leaders can anticipate changes and leverage them to create competitive advantages. Please

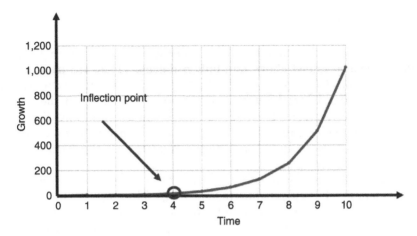

Figure I.1   Inflection point.

note that we are using the exponential curve here for illustrative purposes only and not in its precise mathematical sense. For a more extensive discussion on this subject, please refer to the Appendix.

The key is to detect the signals of change and respond early enough, before it is too late. The challenge is always separating the signals from the noise which is often pervasive and distracting due to variations in the data. Startups have to be very good at detecting inflection points and making distinctions between signal and noise in order to survive. Most companies have a difficult time seeing inflection points and positioning themselves to take advantage of the opportunities these changes represent before they become threats. Technology startups tend to be nimbler, closer to the action, and demonstrably more willing to experiment and change directions than their more established counterparts. In the startup world it is very common to pivot to a different business model if the first one tried does not achieve the expected results. But for many legacy companies, it is a lot harder to pivot, partially because of their commitment to past investments, but in some cases also due to their intrinsic belief that what has worked in the past will also work in the future. Many legacy companies are focused on short-term metrics derived from strategies that yielded results in the past, and their resources are allocated accordingly. However, allocating resources to conduct the necessary explorational experiments intrinsic to growth, but that requires taking some risk, is not common practice in these types of companies.

Trillion-Dollar Platforms

As of this writing, according to CB Insights, there were more than 400 private companies valued above one billion dollars.[7] They are part of the Global Unicorn Club. What is remarkable is that most of these companies are recently formed, technology-based enterprises. They run the entire spectrum of technologies from Security to eCommerce, Internet Services, Artificial Intelligence, and so on. These are the companies that are able to take advantage of the huge opportunities converging technologies in the Exponential Era represent.

And we are just scratching the surface of what is coming. The economic opportunities to be generated by the creation, rapid adoption, and convergence of emerging technologies is truly extraordinary. According to the World Economic Forum there has been no historic precedent to this current phenomenon, which is also referred to as the 4th Industrial Revolution, in terms of its velocity, scope, and impact on everything in our lives.[8] You would have to go back to the late 1800s and early 1900s to see just three significant innovation platforms come together over several decades: electricity, the telephone, and the internal combustion engine.

Today there are at least 10 of these platforms, depending on how you categorize them, that have surfaced in the last few decades: Biotechnology, Nanotechnology, Autonomous Vehicles, Robotics, 3D Printing, Artificial Intelligence, Blockchain Technology, Augmented and Virtual Reality, and the Next Generation Internet. The latter contains several sub-components such as Mobile Payment, Internet of Things (IoT), Online-to-Offline (O2O), and 5G. We will discuss many of these platforms in detail in Chapter 2.

These platforms are generating entirely new ecosystems and multi-trillion-dollar economies. Their convergences are creating a combinatorial power that is spawning ever-increasing innovation and economic opportunities and threats across the globe. For example, today we see China leapfrogging to modern technologies like O2O, machine learning, and mobile payment. The latter has created a cashless ecosystem that has grown to 24 trillion dollars, almost twice the size of China's GDP. When these synergistic technologies with multi-trillion-dollar scope start converging and feeding on each other as we see in China, the explosive scale, scope, and speed of the Exponential Era become frighteningly obvious.[9]

There are many imminent threats to slow-moving companies that insist on operating on outdated business models or are incapable of adapting to fast-moving changes. But there are also enormous opportunities for both incumbents and newly formed enterprises that see the inflection points before they happen. Today's opportunity, if left untapped, can turn into tomorrow's threat, or as we like to say, "the difference between a threat and an opportunity is the time horizon in which you see it."[10]

Changing Wheels While the Car Is Running

Incumbents – the current businesses that are growing profitably in their respective markets – need to be able to continue operating with excellence, maintaining their leadership positions, while concurrently seeking new opportunities in current, adjacent, or brand-new markets. We see them as ambidextrous, able to execute both exploration and exploitation.

These companies are introducing automation, finding additional efficiencies, and seeking to gain additional share in their existing markets while at the same time discovering new opportunities that in many cases are designed to cannibalize currently profitable but eventually poorly performing businesses.

Companies that successfully navigate through the inevitable turbulence in their markets are able to balance the exploitation vs. exploration equation. Their journey is typically marked by a series of consecutive "S-curves," characterized by an initial period of exploration, then growth, and eventually a plateau. For instance, IBM moved through several S-curves in its long history, starting with counting machines and typewriters, then moving into mainframe computing, client-server software, consulting and services, and finally cloud computing (Figure I.2).

Ambidextrous organizations are able to traverse several S-curves, from the plateau of exploitation to the initiation of exploration, because the leadership team can articulate a vision and a set of values of the company that promote a common identity, even if the culture of the exploiting business units is different than that of the exploring ones. These leaders own their ambidextrous strategy and are comfortable designing separate business units that, despite their differences, share common values, and can collectively pursue goals and objectives. They are able to allocate resources and solve conflict, while maintaining alignment with the broader goals of the organization.[11]

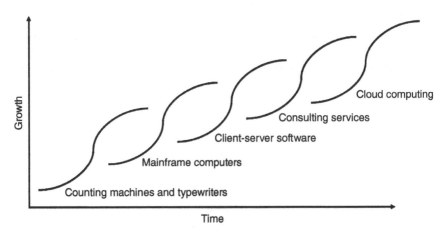

Figure I.2   Series of consecutive S curves.

Dealing with Chaos

So far, we have made the case that we are living in a new era where changes are happening at unprecedented speeds, and where the rapid convergence of technology vectors is creating ecosystems imbued with new threats and opportunities that accelerate with astonishing speed. We have identified the importance of seeing the inflection point in exponential curves before it happens in order to benefit from the resulting changes. We have also recognized that culture plays a critical part in a company's strategy and ability to embrace change, and that the leadership team of incumbents needs to have an ambidextrous strategy, exploiting the existing S-curve, while simultaneously exploring new ones.

This may seem fairly straightforward on the surface, but the reality is that companies struggle to detect the early signals that warn them of an impending inflection point. Even if they see the signals, unless they have a carefully crafted methodology that guides them through their strategic planning processes, chances are they will not take the actions necessary to leverage the opportunities and mitigate the threats of exponential changes.

So how can large companies develop the same capabilities as startups that enable them to detect early signals? How can they "see around corners" as Rita McGrath calls it? How can the leadership team incentivize the organization to pursue new opportunities and create a culture that embraces exponential change, and the chaos that it can create, as a natural part of doing business in this new era?

We believe the answer lies in adopting a robust methodology and a set of tools that the organization can use to guide it through the disruptive nature of the Exponential Era. We call it Strategic Planning for the Exponential Era, or SPX for short.

We will spend a significant portion of the book unpacking the basis of SPX and its fundamental activities. SPX was born out of a set of principles and methods from disciplines as diverse as software development, manufacturing, design, and the military. It applies to strategic planning the benefits that other methods that embrace agility and experimentation have brought to the solution of complex problems when operating in an environment that is uncertain and that changes at fast speeds.

Uncertainty Must Be Managed

We can't promise you that we will eliminate uncertainty. Nobody can. But what we can tell you with a high degree of confidence is that there are robust methodologies and techniques that have been used effectively in diverse disciplines to manage uncertainty and the complexities associated with change. In the software field, where requirements are constantly changing, and the development of complex algorithms requires collaboration from individuals with diverse skills, the Agile movement brings much-needed stability to a discipline that tends to be chaotic. In manufacturing, where rapid changes in customer demands, supplier schedules, engineering configurations, and production capacity challenge the ability to plan, Lean thinking provides innovative ways to address those challenges. In the world of design where there are no fixed solutions and no right or wrong answers, Design Thinking comes to the rescue. In the military, where lives are at stake in the battlefield, and leaders must make decisions under time pressure to outsmart opponents in the fog of war, the methodology developed by the US Air Force for air to air combat – Observe, Orient, Decide, Act – known as the OODA loop, provides a helpful framework to deal with such uncertainties.

As we explored how different disciplines deal with the types of challenges that executives encounter in developing strategies amid uncertainties typical of the Exponential Era, what stood out to us was just how much commonality was found in how these disciplines approach the problem. Our belief, and an important tenet of this book, is that the convergence of these principles, methods, and techniques used in multiple disciplines to deal effectively with the uncertainties of fast-changing environments, connected in unique ways, form the basis of an effective methodology for strategic planning in the Exponential Era.

The Premise Behind SPX

SPX helps organizations identify, explore, monitor, and intercept "horizons" – the changing trends and the technology convergences that can be hugely advantageous to the business if caught early enough, allowing timely and targeted investments, or disastrous if recognized too late, regardless of how much investment is made attempting to catch up.

The premise behind SPX is that it is possible to detect the early signals of transformative disruptions and to map risks, opportunities, and capabilities, allowing the creation and prioritization of actionable plans designed early enough to keep ahead of exponential change. Let's make one thing clear. We are not suggesting that SPX will help companies predict the future. What SPX does is help companies navigate through this new era of swift changes in technology, business, and society. It does this by monitoring the early signals that help anticipate changes and by mapping risks, opportunities, and capabilities, facilitating the prioritization of actionable plans to keep ahead of these changes.

Table I.1   Sample list of products shut down by Google.

Product	Description	Longevity
Google+	Google+ was an Internet-based social network.	2011–2019
Inbox by Gmail	Inbox by Gmail aimed to improve email through several key features.	2015–2019
Picasa	Picasa was an image organizer and image viewer for organizing and editing digital photos.	2002–2015
Orkut	Orkut was a social network designed to help users meet new and old friends and maintain existing relationships.	2004–2014
Google Nexus	Google Nexus was Google's line of flagship Android phones, tablets, and accessories.	2010–2016
Glass OS	Glass OS (Google XE) was a version of Google's Android operating system designed for Google Glass.	2013–2017
Google Chrome apps	Google Chrome apps were hosted or packaged web applications that ran on the Google Chrome browser.	2010–2017
Google hands-free	Google hands-free was a mobile payment system that allowed users to pay their bills using Bluetooth to connect to payment terminals by saying "I'll pay with Google."	2016–2017
Android @ Home	Android @ Home allowed a user's device to discover, connect, and communicate with devices and appliances in the home.	2011–2015

Source: Data from Google Graveyard, Killed by Google.

Organizations that want to not only survive but thrive in the Exponential Era will have to embrace disruptive change and discontinuity. As uncomfortable as this may seem to companies with traditional business models and behaviors, with the right support from the senior management team, and the right shared values and frameworks, organizations can develop behaviors and cultures that embrace change. We see this type of culture in relative newcomers, like Facebook, where employees are told to "break things," or Google, where employees can dedicate a portion of their time to work on projects of their choosing. Google is proud to disclose an extensive list of products and services that they decided to shut down. The website "killedbygoogle.com" lists 190 of these products and services as of this writing, and you can see a sample list in Table I.1. These shutdowns are not considered failures, but instead, they are indispensable experiments that guide Google in its never-ending pursuit of what they refer to as "moon-shots," an attempt to create game-changing products and platforms.

But it is not only newcomers like Facebook and Google that are capable of developing change-centric cultures. As we have discussed, older, more entrenched organizations can also develop such cultures, albeit, in the cases of IBM and Microsoft it required a change at the CEO level.

Experimentation is an integral component of SPX. We believe that in response to fast changes in technology, organizations must implement small projects that are iterative in nature, and that run continuous short cycles. However, experimentation is not sufficient – it is just one aspect of exploration. We also believe that in order to identify business inflection points, organizations need to conduct research long in advance.

By using Artificial Intelligence to uncover new insights from patent databases, companies can identify leading indicators of emerging technologies that may have a future impact on their business. We will cover this topic in detail in Chapter 4 where we discuss the Predictive Analytics component of SPX.

To summarize, SPX provides leaders a clear methodology for innovation in the Exponential Era. Specific tools and guidance are included for a number of critical aspects of innovation: Ideation, as well as deep analysis, prioritization based on both risk and reward, market trends and competitive insights, fast iterations for both prototyping and commercialization, and knowing when to quit and when to stay in the game. This is all built upon a foundation of culture, behaviors, and executive engagement. Our goal is to help organizations embrace a process and mindset that deal effectively with continuous disruption and chaos. The word "continuous" has often been used within the context of continuous improvement and operational efficiency. Operational efficiency is not in the scope of SPX. We are also not interested in addressing incremental improvements. Instead, our focus is on disruptive changes, and the use of the word continuous is to emphasize a distinction between the iterative nature of SPX versus the discrete nature of traditional strategic planning.

How This Book Is Organized

Several books and articles have been written about the Exponential Era, albeit, they may refer to it by a different name, such as the 4th Industrial Revolution. Even though much has been written about the changes that are occurring during this period, we felt that there was a gap in providing business leaders with a robust, well-thought-out, and effective methodology. One that not only can create clarity around these changes but also develops an actionable plan for benefiting from the opportunities and for mitigating the risks unearthed in this era. That became the impetus for our efforts in developing a new methodology for strategic planning and in describing it in this book.

The book is organized into three sections. In Section I we provide the context for the realities of the Exponential Era. We explain why it is so difficult for people to understand the nature of exponential growth. The gradual nature of the early stages of the exponential curve can be deceiving. We struggle to see how quickly exponential growth accelerates once it gets past the inflection point. These inflection points are so difficult to foresee that many large and successful companies have completely missed them, with disastrous results. We describe how Gordon Moore, co-founder of Fairchild Semiconductor and CEO of Intel, published in a 1965 paper his observations about the doubling of the density of integrated circuits every 18–24 months, and his projection that this would continue for decades. The overarching principle behind Moore's observation, that computing capacity would continue to double every couple of years, still holds today, albeit, through unanticipated means of innovative materials and new technological approaches. This exponential growth of computing capacity has been one of the foundational forces driving the Exponential Era.

In Chapter 2, we discuss megatrends that are being shaped by the convergence of technology platforms. We describe several of these technology platforms, including Artificial Intelligence, Blockchain, Internet of Things, Biotechnology, and others. We discuss how the convergence of these platforms is creating trillions of dollars in economic value and generating unforeseen possibilities for business disruption and transformation. One example is how IoT and 5G combined to create a new ecosystem of superfast hyperconnected sensors and devices, allowing us to monitor and proactively deal with a vast number of situations, from weather changes to the spread of diseases.

In Chapter 3, we depict the types of organizations commonly found in the Exponential Era. We tell the story of several companies whose inability to intercept changing horizons led to their demise – we call them Flash Boiled Frogs. We show how emerging companies that took advantage of inflection points quickly grow to billion-dollar valuations – the Disruptive Unicorns. We define Fast and Furious Gazelles, companies of all sizes that are able to grow very fast. We also describe large established companies – the Dancing Elephants – that use ambidextrous capabilities to simultaneously manage exploration and exploitation, jumping from one S-curve to the next. And of course, we could not leave out the Dominating Gorillas, companies that are achieving overwhelming dominance in every sector of the exponential economy.

Section II is about how organizations can deal effectively with the realities of the Exponential Era by following a robust methodology that challenges the current thinking in strategic planning. In Chapter 4, we introduce this new methodology we call SPX and describe how it differs from traditional strategic planning processes. Then we get into the heart of the methodology. In Chapter 5, we explain how to listen to the early signals in order to identify and monitor horizons. In Chapter 6, we illustrate how companies can generate insights by setting up and learning from experiments. In Chapter 7, we discuss how to map organizational capabilities to risks and opportunities. In Chapter 8, we show how companies can formulate and implement plans, and make prioritized, feedback-based, data-driven decisions about strategic initiatives. Chapter 9 is a discussion about leadership, and how to create a culture that embraces change.

We close in Section III with a single chapter. Chapter 10 provides a discussion about the impact of the Exponential Era on humanity, and how the rapid changes we are experiencing challenge our current societal structures, economics, and ethics. We reflect upon our individual and collective legacies and the world we will leave for our children.

SECTION ONE

CHAPTER 1

The New Context for Our Future

According to Dr. Albert Bartlett, Professor Emeritus at the University of Colorado at Boulder, "The greatest shortcoming of the human race is our inability to understand the exponential function."[12] In March 1996, Nicholas Negroponte, head of MIT's Media Lab and author of *Being Digital* expressed the same sentiment less eloquently but in a very straightforward manner: "People don't get exponential."[13] More than 20 years have passed since Negroponte's pronouncement, but that reality still holds true today. To this date, people still do not understand the power of the exponential curve. Perhaps this is a reflection of its deceiving nature, and how the exponential curve manifests itself in a way similar to a line in Hemingway's *The Sun Also Rises*: "Gradually, then suddenly." Or perhaps it is a reflection of the fact that exponential is simply too difficult a concept to grasp for brains evolved to work efficiently based on the timing of circadian rhythms that function best with emotive images, and that prefer to think linearly.[14]

Our continued use of agrarian words to describe business activities, like "seed" (as in seed money in the venture world), "plant" (as in initiate a foothold in a market), "cultivate" (business development of a new market), "harvest" (sell), and "cash cows" (usually revenue streams that are being "milked" for remaining profits before the market collapses or is disrupted) has locked our thinking into a pattern of behaviors completely out of sync with our era. In the Exponential Era, we don't operate on circadian rhythms or agrarian time scales.

This is an era marked by the confluence of fast-changing technologies that converge to create new ecosystems, resulting in digital disruptions at a rapid velocity. It ignores our hard-wired primal core, leaving us slow to adapt

The Exponential Era: Strategies to Stay Ahead of the Curve in an Era of Chaotic Changes and Disruptive Forces, First Edition. David Espindola and Michael W. Wright.
© 2021 by The Institute of Electrical and Electronics Engineers, Inc.
Published 2021 by John Wiley & Sons, Inc.

to changes that are happening in new time scales. The technology growth that we are experiencing today does not follow linear progressions like animal migrations, growing seasons, and calculated production runs.

In the Exponential Era, having access to data and having the ability to understand the information the data is telling us is on every individual's and organization's critical path for survival. It is an era where "knowing" is becoming paramount to surviving. The traditional "have's and have not's," which referred to one's means, have accelerated, transformed, and become "those who know and those who don't."[15] This is an era that runs on creating, harnessing, intercepting, and integrating technologies at speeds and scales never before experienced in human history.

The Future Is Not Your Brain's Priority

Our inability to relate to the Exponential Era is most pronounced when we try to grasp the fundamental construct of exponential curves. Picturing in our minds how technologies move imperceptibly across what appears to the brain as distant and slowly approaching horizons – only to be surprised as they suddenly explode in front of us at speed and volume – is one of the great mental conundrums of the Exponential Era. These explosions in the growth and the unprecedented rates of adoption of new technologies are leaving most of us unprepared and in wonder. We find ourselves trying to set our minds to a view that is capable of constantly adapting to the sudden appearance of new technologies and digitally driven transformations.

These digital transformations are the direct result of the confluence of new technologies converging to create entirely new ecosystems.[16] These new ecosystems grow at velocities well beyond our primal brain's hunting mode speeds. Calculating where to launch a spear and at what speed to intercept a target remains a difficult task that takes time and practice to master. However, a computer aims, calculates range, fires, and hits a target in milliseconds, again and again.

Humans have difficulty observing and responding to the future. We have trouble extrapolating meaning or even putting energy into understanding a time period that appears to be far away. While some of us can marshal and focus our brain's processing energy on "futures," most of us can't. In fact, very few of us can focus for extended periods on our future. The reason is fairly simple. Our brains are not comfortable with diverting energy from human self-preservation and survival. In order to survive, we are careful with how we allocate our finite brain energy; and it is very much "in the moment!"

Research shows that our brains think that concentrating on our current self is rather more important than worrying about our future self, let alone future generations.[17] The study of fMRIs overwhelmingly concludes that the energy that our brains put to our current self, relegates the future self to a much lower priority. As Jane McGonigal, director of the Institute for the Future writes: "Your brain acts as if your future self is someone you don't know very well and, frankly, someone you don't care about."[18] It seems only logical. After all, we have survived as a species for a long time by being alert to immediate threats. Our brain activity is largely occupied with operating life-preserving processes and looking for threats and pleasures right now. This leaves us wide open to dramatically underestimating the real impact of exponential curves over time.

Exponentials Make Fools of Humans

Humans are easily fooled by exponentials. To illustrate the point, we will borrow from the famous fable of the origin of the game of chess. Legend has it that the creator of chess presented it as a gift to an emperor in India. The emperor was so impressed with the ingenuity of the game that he felt compelled to compensate the man and asked him what he would like to receive as his reward. The man humbly responded:

> "Oh emperor, my wishes are simple. I only wish for this. Give me one grain of rice for the first square of the chessboard, two grains for the next square, four for the next, eight for the next, and so on for all 64 squares, with each square having double the number of grains as the square before."

The emperor was astounded that this ingenious man only wanted a few grains of rice as his reward for such a wonderful game. Without much thought, he granted the man his wishes. It wasn't until sometime later that his treasurer came back and advised the emperor that it would be impossible to pay the man the quantity requested, as the grains of rice added up to an exorbitant amount: 18 quintillion (18 followed by 18 zeroes) to be exact, the equivalent of roughly today's entire worldwide crop for a decade.[19]

How could the emperor be so easily deceived? Simple. He was thinking linearly like most of us often do, while the ingenious man understood and used the power of the exponential curve. And herein lies the threat, or the opportunity, depending on your point of view. Humans tend to think linearly, but the technology changes we are experiencing now follow an exponential curve.

The Phases of Exponential Progression

We introduced you to the concept of the inflection point in the beginning of the book. Some authors refer to it as the elbow of the exponential curve. When exponential growth is plotted on a graph, it appears as if very little, if anything,

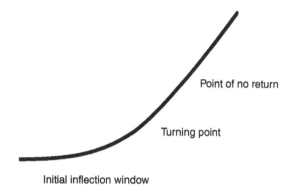

Point of no return

Turning point

Initial inflection window

Figure 1.1   A closer look at the elbow of the exponential curve

is happening for a long time. The graph is essentially flat until suddenly an elbow appears. At first glance, that elbow may seem like a single point on the graph where the inflection happens, and the curve starts to go way up. But in reality, the business characteristics of an inflection have three distinct points, as illustrated in Figure 1.1: the Initial Inflection Window, the Turning Point, and the Point of No Return.

These distinct points mark important phases of growth that have significant implications for the development of business strategies. These phases are analogous to what you might encounter in Gartner's hype cycle, to represent different stages of technology development and commercialization.[20]

Before the Initial Inflection Window, as discussed before, the graph is basically flat. During this period, new technologies are being developed, new terms are being coined, and the press is starting to pay attention to these developments. Think of artificial intelligence in the 1950s or Bitcoin in the mid-2010s. This may lead to some initial "hype," and you may start to hear pundits proclaiming that a new world order has arrived. Inevitably the initial hype leads to disillusionments as these technological wonders fail to deliver on their overhyped powers.

This is when you will start to hear the naysayers remind you how they were right, and how this overhyped change is never going to happen. At this point, only a few initial entrants to the market will have survived as the Initial Inflection Window reveals itself. This is the time period when threats and opportunities become more pronounced.

It is during the Initial Inflection Window that the early signals start getting stronger and clearer. At this stage, it would be advisable to set up several short experiments to explore the possibilities and start putting the proverbial "toe in the water." This is the time to make initial small investments in capabilities, but it is not yet the time to make big bets.

Subsequently, the inflection point starts to become more pronounced and as those early market entrants start to get traction, we enter the Turning Point. By this stage at least a rough plan should be in place based on assessments of risks and opportunities. A priority list based on early experiments will guide investments as the organization iterates on gathering data, collecting market feedback, and making executive decisions on what initiative to scale or terminate. We will cover this process in detail later in the book.

Eventually the growth accelerates as we get past the Turning Point. At this stage, the benefits created by the exponential growth have become apparent to everyone, and what used to be an opportunity has turned into a threat for those who adopted the wait-and-see approach. Those who did not engage in the previous stages of the inflection point or who stubbornly stuck to their now stale "strategic plans" scramble to catch up watching powerlessly and in disbelief as their businesses go into decline. They have now reached the Point of No Return.

At this stage, what seemed like a faraway future has become the new reality, and it bears no resemblance to the past. It may appear as if it happened suddenly, almost magically, like a marvel that came out of nowhere. But in reality, the future does not just happen all at once, as McGrath insightfully noted. It begins to unfold in a cadence that allows you to detect early signals, but only if you are paying attention.

Humans tend to ignore the early inklings, lose patience with the slow progression of change during the early compounding, miss the lift-off in the inflection window, chase past the Turning Point, and then become exasperated when the curve disappears into the Point of No Return. All of it in what seems like the blink of an eye.

In addition to our natural struggles in seeing exponential change, hierarchies in large organizations also create dangerous blind spots. The early evidence of change seldom reaches the CEO's office or the corporate boardroom. In fact, we know that by the time information filters through several layers of management, the early signals that are so critical to spotting inflection points in the Exponential Era are muted by PowerPoint presentations that lack granularity and clarity. The early signals are usually only visible by those working at the frontlines. It is the frontline employees who first hear about competing startups that are hitting the market; disrupting technologies that are threatening existing business processes; or customers who are migrating to cheaper, faster, and more efficient ways to get what they need. As Andrew Grove eloquently explained in *Only the Paranoid Survive*, "When spring comes, snow melts first at the periphery, because that is where it is most exposed."[21]

Intuitively, being alert to the early signals that may indicate a potential business inflection point makes perfect sense. Yet, it is surprising to see the number of senior executives whose agendas are filled with urgent matters related to making this quarter's numbers. Instead of being at the periphery, paying attention to the early signals, many spend their time in internal meetings, surrounded by team members who are worried about self-preservation and don't necessarily want to "rock the boat."

The danger here is that you may end up building a self-reinforcing echo chamber that filters any signals contrary to deep-rooted beliefs. Leaders may have the best of intentions as they communicate their visions and encourage their troops to go out and win more business. Middle managers will "drink the kool-aid" but will avoid reporting conflicting news up the chain. Frontline employees may unintentionally hide information from their managers in a misplaced effort to give a good impression. Consequently, senior executives become completely isolated from the realities and the signals coming from the edges. They may think that their vision and strategies are spot on because they never get to hear conflicting signals.

Amy Webb, Founder of the Future Today Institute and author of the bestseller *The Signals Are Talking: Why Today's Fringe Is Tomorrow's Mainstream*, is a great proponent of creating mechanisms that allow organizations to effectively forecast changes so that they can identify risks and opportunities before disruption hits.[22] These mechanisms force the organization to be on the constant lookout for signals that are not visible through the usual channels. They bypass organizational hierarchies and introduce processes in which there is a continuous cadence of discoveries and learning opportunities across all layers of the organization, from individual contributors to the CEO.

Such mechanisms are implemented effectively by organizations that have adopted a forward-thinking, discovery-based methodology and whose leaders have instilled a culture that embraces changes and risk-taking. This is the essence of what we are going to explore with you in this book.

> "A strategic inflection point is a time in the life of a business when its fundamentals are about to change. That change can mean an opportunity to rise to new heights. But it may just as likely signal the beginning of the end."
>
> ANDREW GROVE

The Exponential Explosion

As we explore fast-growing technologies that are impacting our future, consider the following exponential curves, their accelerated growth, and consequent cost reductions.

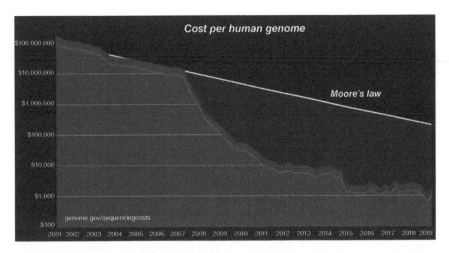

Figure 1.2   Cost per human genome
Source: NIH. National Genome Research Institute. Public Domain. https://www.genome. gov/about-genomics/fact-sheets/DNA-Sequencing-Costs-Data

In telecommunications we find the following remarkable advancements:

- Fixed broadband speeds will more than double by 2023.
- Mobile (cellular) speeds will more than triple by 2023.[23]

In biotechnology, the costs to sequence a base pair of genes have decreased by half every 18 months for the last decade (Figure 1.2).

Note that there are approximately 1,000 nucleotide pairs of coding sequence per gene. The human genome contains approximately 3 billion of these base pairs. Thirty years ago, a scientist could expect to spend 20 years exploring the sequence of one gene pair. The first whole human genome sequencing cost roughly $2.7 billion in 2003. By 2006, the cost had decreased to $300,000. In 2016, the cost reached a remarkable and generally affordable $1,000 mark and now a Chinese company claims it can sequence the human genome for $100.[24]

Another remarkable area of exponential growth is the amount of knowledge humans have accumulated in less than a century. If you were born in the late 1940s to early 1960s (aka baby boomer), while you were growing up you wouldn't have seen the beginning of the exponential growth of data accumulation and the expansion of knowledge. But by the time you reached adulthood it was changing rapidly enough for you to notice the inflection point. If you are a young adult today, you are experiencing first-hand the overwhelming growth and impact of knowledge expansion at an exponential rate. IBM predicts that knowledge will double every 11–12 hours in 2020.[25]

Moore's Law: The Observation that Changed the World

If the progression of the numbers 32, 64, 128, 256, 512, 1,024 looks familiar to you, it is because you have seen them on your camera memory discs, your computer, your smartphone, and your TV. The fact that specifications for such devices have been doubling is because of an exponential technology called semiconductors. If you have experience in the semiconductor market you will recognize the number progression as nodes along Moore's law.

Moore's law is a prediction made by Gordon Moore in 1965 that the number of transistors per silicon chip would double every 18–24 months. Stated another way, Moore predicted that the size and cost of the devices in our digital world would halve and their computing power would double about every two years.

The impact of Moore's law in the increase in the density of integrated circuit chips is illustrated in Figure 1.3. This increase drives computing power.

Today, the world's fastest supercomputers can perform 1×10^{18} calculations per second, otherwise known as exaFLOPS. They can handle an astonishing amount of data, with bandwidth 24 million times greater than the average home internet connection, and capable of processing 100,000 HD movies in a second.

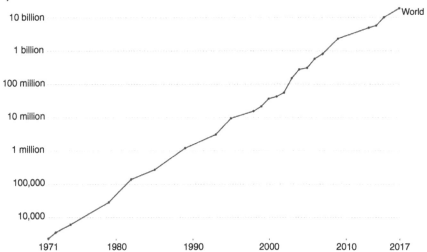

Moore's law: transistors per microprocessor

Number of transistors which fit into a microprocessor. This relationship was famously related to Moore's law, which was the observation that the number of transistors in a dense integrated circuit doubles approximately every two years.

Figure 1.3   Moore's law: transistors per microprocessor
Source: Our World in Data. "Moore's Law: Transistors per Microprocessors." https://ourworldindata.org/grapher/transistors-per-microprocessor. Licensed under CC BY 4.0

Along with processing speed, as we point out above, the density of storage capacity is also increasing. In just 50 years we have experienced a 25,600 times reduction in size for the same amount of storage. It is the equivalent of taking seven football fields down to the size of a quarter.

Over the first 50 years of Moore's law, we have seen the convergences of technologies create the early stages of exponential change in multiple industries and applications. It is not just a single exponential curve that is creating disruption, but the confluence of interconnected technologies that grow at exponential rates. As digital technologies influence the advancement of other technologies, the exponential nature of Moore's law reaches broader and more impactful significance.

Consider, for example, how global connectivity and the miniaturization of computing devices have converged to revolutionize communications and enable commerce anytime, anywhere. The impact has been nothing short of astounding. We carry in our pockets computers that have a million times more memory than the computers that guided us to the moon.[26] We can communicate instantaneously with more than half of the world's inhabitants who now have access to smartphones. Small farmers in Africa can now check market prices for their crops instantaneously, allowing them to better negotiate prices and offer their products to the highest bidders. Our cars can park themselves and soon will be driving on our highways without human intervention or supervision. The entire taxi industry has already been disrupted, and now leading logistics and transportation companies are transforming themselves to become digital behemoths.

What just a few years ago was only possible in the imagination of Hollywood producers is now within our grasp. Neuroscience converging with computer technology now gives us the ability to control objects with the power of our thoughts. The CEO of a Brazilian nonprofit organization has become the first person to drive a Formula 1 racing car using only the power of his mind. Rodrigo Hübner Mendes, a quadriplegic, and the Founder and CEO of the Rodrigo Mendes Institute, used Brain–Computer Interface (BCI) technology, which was developed by Tan Le, Founder of EMOTIV Inc., to pilot the vehicle by thought alone. "Last month I went to a speedway in Brazil and I had the opportunity to drive a racecar using my mind," Mendes said. "The car, it doesn't have pedals, it doesn't have a steering wheel, it doesn't have anything – it's just him and his mind, driving it forward. It blew my mind," Le explained.[27]

Each new ecosystem created by the convergence of technologies is producing entirely new ways of delivering rewarding experiences and changing behaviors like never before in human history. The exponential era is changing our lives, our businesses, our economy, how our societies function, and our worldviews on civilization.

The future of work is being redefined by the convergence of artificial intelligence and robotics. According to a 2018 McKinsey Global Institute report, it is estimated that as many as 357 million people around the world may need to change their occupations and acquire new skills.[28] The socioeconomic implications are dumbfounding.

The adoption of new technologies at exponential rates is also creating new ethical challenges and the need to develop new ethics to deal with technology's restless and unstoppable search for "progress." We simply have not proven capable of seeing the results of our actions over time. Our current ethics are based on what can be observed. However, the totality of the consequences of our technologies is no longer observable before launch and adoption.

We hope we have impressed upon you that exponential growth is not so obvious in the beginning, but that it has a dramatic finish. The result of exponential growth sometimes has positive consequences, and sometimes not. Therefore, we must reflect upon how we deal with ethics in this era of unprecedented technological, environmental, economic, and societal changes. We will cover this subject in detail in Chapter 10.

In the next chapter, we discuss megatrends shaped by the convergence of several technology platforms that are creating trillions of dollars in economic value. We discuss how these platforms, such as Artificial Intelligence, Blockchain, Internet of Things, Biotechnology, and others are generating new ecosystems and creating unforeseen possibilities for business disruption and wide-ranging transformations.

CHAPTER 2

Exponential Platforms: Convergences and Megatrends

What you are about to read may all seem like science fiction, but nothing that we portray here requires a new technology that has not yet been invented or that is not operational. It is only a question of time and improving access and distribution to achieve the pervasiveness of the trends we describe here. In every era of change, transition periods arise that are disruptive, chaotic, and transformative. The Exponential Era is no exception. The convergences and megatrends outlined here will happen. Their timing and pervasiveness will be influenced by unforeseeable accelerants and inhibitors.

So far, we have presented the idea that the exponential growth of technology is driving unprecedented changes to all aspects of our existence. Having done that, we believe it is also important to acknowledge that not all technologies grow exponentially, and that technological growth is influenced by many factors such as legal, economic, environmental, social, and political considerations. The interaction of all these factors combined is what determines to what degree we experience change and adoption.

However, it is becoming increasingly clear that what Ray Kurzweil, founder of Singularity University, discovered in the 1990s and called the "Law of Accelerating Returns" is proving to be true. Once a technology becomes digital, it starts to gain the growth characteristics of Moore's law and begins to accelerate exponentially. The most formidable innovations of the twenty-first century, including Artificial Intelligence (AI), Robotics, Nanotechnology, 3D printing, Blockchain, Augmented Reality (AR), and many more, are accelerating at this rate. We refer to these innovations as exponential platforms due to their growth rates and scale.

The Exponential Era: Strategies to Stay Ahead of the Curve in an Era of Chaotic Changes and Disruptive Forces, First Edition. David Espindola and Michael W. Wright.
© 2021 by The Institute of Electrical and Electronics Engineers, Inc.
Published 2021 by John Wiley & Sons, Inc.

Exponential platforms and their potential impact on society are astronomical. When these platforms converge, they have the potential to unleash unfathomable changes that are difficult for even the most astute futurists to imagine. The convergence of exponential technology platforms is enabling a new scale of disruption that we have never experienced before. These disruptions are driving megatrends and creating new ecosystems that will have a profound impact on every aspect of human existence.

In addition, we live in a hyperconnected world with synchronous communications. If something happens in one corner of the world, we hear about it in seconds. Our economies are interconnected in such a way that an event in one country or region has repercussions throughout the globe. Infectious diseases, as we have seen with the coronavirus, spread exponentially to all areas of the planet in a matter of weeks due to the large number of people who travel and have contact with others across the globe.

The combination of exponential technology growth, convergence, and globalization will bring about changes of such scale and velocity that our linear thinking brains can hardly comprehend, challenging our ability to cope and adapt. According to Kurzweil's Law of Accelerating Returns, we are going to experience 20,000 years of technological change over the next 100 years, the equivalent of going from the dawn of agriculture to the invention of the internet twice in one century. Peter Diamandis, in his book *The Future is Faster Than You Think*, describes it this way: "This means paradigm-shifting, game-changing, nothing-is-ever-the-same-again breakthroughs will not be an occasional affair. They will be happening all the time."[29] As discussed before, our brains have a limited ability to project into the future, and that may explain why you may experience difficulty processing and internalizing the discussions in this chapter and consider the magnitude of these changes exaggerated.

The convergence of dozens of exponential platforms will spawn ecosystems that will result in ever-increasing changes to how we live and interact with each other as human beings, creating social, economic, and political ramifications of unprecedented nature. These ecosystems are expected to create instantaneous access to information, more time for leisure, higher life expectancy, and greater opportunity for wealth creation. Diamandis states that "Every time a technology goes exponential, we find an internet-sized opportunity tucked inside. . . As a result, if the internet is our benchmark, more wealth could be created over the next ten years than was over the previous century." He adds further: "Unicorn formation, or the time it takes from 'I've got an idea' to 'I run a billion-dollar company' was once a two-decade long shot. Today, in some cases, it's nothing more than a one-year adventure."

Our ability to adapt to this new normal, both as individuals, as well as institutions, will be put to the test. Many companies will have a hard time keeping pace. According to Richard Foster from Yale University, 40% of today's Fortune 500 companies will be gone in 10 years.[30] Our eighteenth-century based education institutions will have a difficult time justifying the currently unsustainable cost of higher education that has led to the student debt crisis, and many will not survive. We believe the migration to a lower cost distance-learning model which has been long in the making will be accelerated by an unexpected catalyst: social distancing imposed by the coronavirus.

Our ability to retrain millions of workers regularly will be paramount to effectively adapt to a world driven by AI and automation. This will also drive changes to the current education model which needs to keep up with swift changes in the skills required to be productive in a whole new category of jobs that don't even exist today.

"We are drowning in information but starved for knowledge."

JOHN NAISBITT

Megatrends

The convergence of exponential platforms will shape megatrends that will bring profound changes to the way we live, how our businesses operate, and the new ways our societies will function. Futurists and analyst firms have identified several of these megatrends. Keep in mind that not all of the forecasted changes will come to pass, but many of them will, with extraordinary repercussions. From the perspective of a point in time in early 2020, we have selected 10 megatrends we believe are going to persist and be impactful in the not too distant future.

Hyper Global Connectivity

We can feel how the world is hyperconnected as news across the globe is instantaneously shared via multiple synchronous communications channels. Applications like WhatsApp, TikTok, Instagram, Facebook, and Zoom have democratized the ability to communicate with anyone, anytime, and anywhere, essentially for free. Social media permits the broad dissemination of ideas across borders in powerful ways. On the positive side, this allows for diversification, democracy, and innovation. On the other hand, these channels can also be used in frightful ways, threatening our privacy and freedom. Fake news, perception manipulations, and other mind-control techniques are being used in unanticipated ways and are a cause for grave concern.[31]

Hyper global connectivity will increase dramatically as we add the rest of humanity to the global network. It is expected that nearly 8 billion people will be online soon, an increase of 3 billion people to the estimated 5 billion connected today, adding tens of trillions of dollars to the global economy. The deployment of 5G technologies and the launch of a multitude of global satellite networks from companies like OneWeb, Starlink, Amazon, SpaceX, and others, will result in global gigabit connectivity speeds at very low cost. SpaceX alone is planning on launching 12,000 satellites, enough to cover the entire planet. 5G technology will permit downloading a high definition movie almost instantaneously to be viewed just about anywhere on the planet.

As with all technology adoptions, but especially those done at speed, the potential for abuse always exists. We have seen and will continue to see cybersecurity breaches, fake news, and other less productive uses of human connectivity. However, we believe that in the Exponential Era, people, companies, industries, and nations will learn from experience and adjust their behaviors accordingly. What cannot be overlooked is that the addition of 3 billion minds interconnected at vast speed will spur innovation at scales never experienced before.

A Longer and Healthier Life

The introduction of several biotech innovations being made available to consumers this decade, such as senolytic medicines and stem cells, is expected to increase life expectancy by 10 years or more. Just to provide a perspective on the speed of growth of human lifespan, in the Middle Ages the average lifespan was 31 years. By the end of the nineteenth century, it broke 40 for the first time. By the end of the millennium, it was 76. Since then we have made continuous progress in detecting and treating the two most fatal illnesses, cancer and cardiac disease, extending today's life expectancy into our 80s. Research shows that as we tackle neurodegenerative disease, we can reach 100 and beyond. Ray Kurzweil believes that we will soon reach what he calls "longevity escape velocity," the point in which we extend life expectancy by more than a year each year: "It is likely [that we're] just another 10 to 12 years away from the point that the general public will hit longevity escape velocity," he says.[32] Interestingly, recent data in the United States shows life expectancy declining slightly year over year due to suicides and drug overdose. These are some examples of the inhibitors that may add noise to the movement of a trend.[33]

It is not just lifespan that will improve, but also the quality of life. Gene editing technologies will revolutionize medical care, allowing personalized treatment and the ability to cure hundreds of genetic diseases. CRISPR 2.0, a next-generation gene editor, allows the change of a single letter in a string of DNA, opening the door to the potential treatment of thousands of diseases.[34] Stem cells, which have the ability to turn into any other type of cell, are one of the body's main repair mechanisms. Stem cell therapy is an area of research with promising treatments for dozens of ailments that afflict humans today.

A focus on prevention will transform medicine in unimaginable ways. We are already experiencing the use of sensors that provide electrocardiogram information from smart watches. With the advance of nanotechnology, we will be able to sense, store, and transmit data from our bloodstream, and instantaneously feed healthcare providers with a dashboard containing all of our vital signs.

Telemedicine will be widely available and effective across the world. In this case, COVID-19 has been an example of an unforeseeable accelerant driving faster adoption of a new technology. Doctors, aided by AI, and with a full picture of your vital signs transmitted by nanobot sensors circulating in your bloodstream, will be able to diagnose treatments more accurately and more effectively without seeing you in person. All of this will be done instantly. Someday soon, house calls will come back, but in a virtual format. No more waiting at the doctor's office!

Augmented and Virtual Worlds

If you have ever played a Virtual Reality (VR) game, you certainly understand where this technology is going. Staring down a precipice wearing VR goggles is just as scary as the real thing, even though your conscious mind knows your two feet are safely planted on firm ground. The Pokémon GO app and the popular YouTube video showing a killer whale jumping in a gym opened our minds to the possibilities of Augmented Reality (AR).[35]

Now every sense is being added to the virtual experience, including touch with haptic gloves, vests, and full-body suits, as well as scent emitters and taste simulators. VR has shown to be more effective in changing behaviors than any other traditional media, creating new possibilities for general education, corporate training, sports, mental health treatment, and numerous other applications.[36]

VR and AR promise to impact every industry from retail and advertising to education and entertainment. Avatars will represent our online personas for work, play, shopping, sports, social interactions, and much more. Multiple parallel economies will unfold where digital objects and digital currency will have real value that can be exchanged in the real world.

High-resolution lightweight VR headsets will transform shopping, allowing us to try on clothing virtually from the comfort of our homes. AI-based systems that know our body measurements will help us find the perfect fit. Our avatars will try on anything we want to buy virtually just as if we were at the store. We'll be able to see how a piece of furniture will fit with the rest of our home décor and experience the changes to the "feel" of the space before a purchase is made. Shopping will never be the same again.[37]

There is always the potential for unintended consequences in the process of adoption of these technologies. Examples might include attempts at mind control, behavior manipulations, and even physical injury due to disorientation. However, we are optimistic that the vast majority of use cases will bring many benefits and further accelerate the adoption of these platforms at volume and velocity.

Abundance

We are on our way to enjoying a global abundance of material needs including energy, food, shelter, and capital. The global middle class is continuing to rise, and extreme poverty will likely continue to drop. The digitization of finance, education, entertainment, and much more, combined with automation and cheap energy will drive down costs, raising the standard of living of those in the lowest rungs of society. Capital availability is expected to continue rising, funding businesses of all types. Crowdfunding will continue to provide easier access for entrepreneurs across the globe to fund their ideas, accelerating innovation.

Large scale 3D printing will allow the creation of shelter at a very low cost. Brett Hagler formed a nonprofit called New Story that designed a solar-powered 3D printer capable of creating 400–800 square-foot homes for just $6,000–$10,000.[38]

Solar, wind, geothermal, hydroelectric, and nuclear power generation continue to advance humanity toward cheap and globally abundant renewable energy. According to Peter Diamandis, the price per kilowatt-hour will drop below 1 cent, and storage will be available below 3 cents per kilowatt-hour, providing affordable energy abundance to the entire world.

Vertical farming, cellular agriculture, biotechnology, robotics, and several other advances in the production of food have the potential to increase productivity and eliminate the biggest food challenge today: distribution. Streamlining distribution channels has the potential to reduce and potentially eliminate food insecurity where it is currently rampant.

Cellular agriculture will afford the production of protein at scales and cost never achieved before. An abundance of higher nutritional content that is cheaper and healthier will further enhance the overall health and quality of life throughout the world.

Access to abundance will force us to rethink our concepts of society, politics, the economy, and businesses, creating a new paradigm shift that will hopefully result in a secure and sustainable future. As we try to balance our enthusiasm for the positive outcomes in all of these megatrends, we are also aware of the implementation challenges with each one. For example, it is possible that we may create "digital deserts" or "digital bubbles." In the early stages of adoption and during the transition process, access to abundance may not be evenly distributed, exacerbating inequality gaps around the world. We need to continuously seek to improve access to each of these megatrends in order to mitigate the dislocations associated with disruption.

Upgraded Transportation

We are already experiencing a deep transformation in transportation. Car-as-a-service started with the concept of utilizing existing capacity in the form of private cars that sit idle most of the time and people who are willing to make extra money by driving, and connecting this supply with the demand of people who need transportation that is more efficient than traditional taxis. This was done via a simple and effective user interface utilizing existing converging technologies such as GPS, mobile networks, and smartphones. It only took 14 months for companies like Uber and Lyft to completely disrupt the taxi industry. And this is just the beginning.

Autonomous vehicles are expected to be fully operational in cities across the world in the next decade. This means that one of America's most beloved material possession, pastime, and status symbol, the car, at least in its current embodiment, may become a thing of the past. Young people will someday have the option of not having to learn how to drive.

Over 1 billion dollars has been invested in at least 25 flying car companies.[39] Large established companies like Boeing, Airbus, Embraer, and Bell Helicopter are also investing in this space. Uber Air, the company's flying car offshoot is planning on having fully operational aerial ridesharing capabilities by 2023. The company plans on bringing the cost of aerial ridesharing to 44 cents/mile; cheaper than the cost of driving.

Virgin Hyperloop One, a company founded by industry luminaries like Elon Musk, Peter Diamandis, Jon Lonsdale, David Sacks, and others, promises to connect Chicago to DC in 35 minutes. According to Josh Giegel, its Chief Technology Officer, "Hyperloop is targeting certification in 2023. By 2025, the company plans to have multiple projects under construction and running initial passenger testing."

Elon Musk's SpaceX megarocket Starship, which travels at 17,500 mph, promises to transport passengers from New York to Shanghai in 39 minutes. According to Musk, this technology will be ready for demo in three years. He admits that it will take a while to get the safety right, but it may be feasible that in the next decade or so intercontinental travel will be an under one-hour affair.

Transportation standards take a long time to implement and like all of our megatrends, exact timing is an unknown. Land use, local licensing, new corridors, service areas, etc. will be challenges to overcome. However, transportation, as we know it, will never be the same. The implications for businesses in the transportation industry and beyond are massive.

Instant Personalization

We have already seen companies like Amazon drive delivery times from days to hours. With the advance of drone and robotic delivery, the last-mile delivery challenge is being further reduced, allowing the delivery of products from local supply depots to our doorstep in record time.

On-demand digital manufacturing through 3D printing will allow personalized products to be manufactured en route and at point of use. Patents have already been filed for autonomous vehicles with 3D printing capabilities. This means that the products that you order may be manufactured during transportation and may be completed just in time when the delivery vehicle arrives at the local distribution center or even directly at your home.

AI will know all our preferences, body measurements, recent purchases, shortages, needs, conversations, facial expressions, and much more. As frightening as this may be — innovators and policymakers will need to be watchful for the prospect of unethical uses of AI — on the positive side we will be able to get personalized services and products that are individually designed for us.

Healthcare will be personalized based on our individual needs and according to our DNA. Individualized medication will be highly targeted to address ailments without the side effects common in today's one-size-fits-all approach to medications. This will challenge our ethics and culture toward disease, heredity, and care classifications including patient profiling, medicine restrictions, and compulsory regimens. None of these trends are free of potentially negative impact, but they will get traction, morph to the needs of the customer, and incur accelerated adoption.

Digitization of Work

Humans have successfully adapted to new forms of work and economic systems throughout history. We have transitioned from hunter-gatherers to an agriculture-based society, to the industrial revolution, to the services and knowledge economy. None of these changes were without pain, but eventually we adapted and improved. Now comes the biggest, most impactful transition of all, the digitization of work. Work will never be the same again.

Most physical objects will have a digital twin, and most fabrication will be accomplished through 3D printing. Transportation will be achieved through autonomous vehicles, both terrestrial and aerial. AI will become entrenched in every aspect of life and work, and will communicate and negotiate with other AIs. It will potentially provide many of the cognitive capabilities needed, including creative tasks, idea generation, and innovation. It will even program and create new, smarter AIs. Google is already using AI today to design chips that accelerate AI.[40]

Most likely humans will engage in collaborative work with AIs, but we don't know exactly how and to what extent. The truth is that we don't even know what work will look like in a decade or two. But the early signs help us paint a faint picture of what is yet to come. Intelligent robots aided by AI will not only do most of the physical labor but also most of the service and knowledge work available today. Will humans still have a role to play in this world of automation and digitization? It is possible that humans will interact and collaborate with the robots and AI, and that we will find a niche role to play. But it is also very likely that the jobs available to humans will be much fewer than they are today.

The obvious question this raises is: how will we sustain ourselves without work? We believe that the demonetization and democratization of goods and services produced, in a world with the potential for sustainable abundance, can address the material needs of most humans. However, some important questions will have to be carefully considered in this new world. One is regarding distribution. The other is regarding purpose. The former will require new social-political-economic frameworks that are yet to be determined. The battle between human greed and compassion will continue to play out no matter what new set of circumstances we find ourselves in. The latter is about finding our purpose in a world that does not involve traditionally defined work and where practically all of our material needs are provided for. There are deeply philosophical and spiritual considerations that will become increasingly more prevalent in this new world. We will discuss this topic in more detail in Chapter 10.

Automated Shopping

As AI becomes deep-rooted in all aspects of our lives, it will know more about ourselves than we do. This means that AI will understand our needs and we will increasingly trust and rely on it to make most of our buying decisions. AI will know our past consuming habits, and exactly what we like. Sensors will detect what needs replacement in our cupboards and refrigerators. Alexa, Siri, Google, or some other intelligent personal assistant will listen, understand, and take action based on conversations that we have.

In other words, shopping, if you have internet connectivity, will begin to be simply an automated process that happens behind the scenes and that will not require much of our attention. Most of our shopping needs will be taken care of without us even realizing it — delivered almost instantaneously and personalized for our taste and specific needs.

All of these incredible benefits will have to be balanced with the potential for narrowing our choices and diminishing our desire to explore. In order to accelerate the ubiquitous adoption of these functions, regulations and oversight may need to be developed, access to smart devices and broadband connectivity made affordable, and higher educational levels provided.

Radical Interfaces

When Netscape gave us the first browser, it radically simplified the interface to the internet. Initially, the World Wide Web was simply a set of static pages that provided a one-way interface. Over time, with the advent of Web2.0, we enjoyed a much deeper experience including social media, wikis, blogs, and eCommerce.

Now we are transitioning to the Spatial Web, enabled by billions of connected devices and interfaced via VR and AR. Powered by a backbone of Internet of Things (IoT) sensors, natural language search, data mining, machine learning, AI, and decentralized blockchain authentication, the Spatial Web offers a new radical interface, unlike anything we have experienced before.

Our physical world will have a digital twin integrated seamlessly and interfaced via wearables, smart lenses, and all sorts of VR and AR devices. The Spatial Web provides a new radical interface with digital enhanced versions of our physical world, blending virtual and real in a way that allows for simulations, parallel economies, and even new political systems.[41]

If that is not enough to blow your mind, imagine having your brain connected directly to the cloud. Ray Kurzweil predicts that this will happen in the mid-2030s. Brain–Computer Interfaces (BCI) is already a reality today. As described in Chapter 1, it is possible today to drive a Formula 1 car with nothing but our thoughts. BCI researchers are working on supplementing cognitive abilities that will enhance our sensory capabilities, memory, and intelligence. This means we may be able to communicate instantly and seamlessly with others across the globe with just our thoughts and tap into new sources of cognitive capabilities. Despite potential benefits, connecting our minds directly to a computer network may be a cause of grave concern to many people.

Sustainable Environment

We are finally realizing the significant damage our choices and decisions have caused to the environment. This increased level of awareness is creating social and political pressure that is driving companies toward investing in sustainability. The pursuit of activities that cause harm to the environment is not only bad for public relations, but it is also simply not a sustainable business practice anymore. A new generation of conscious consumers will vote with their dollars and boycott all types of waste and environmental contamination.

Breakthroughs in material science, supported by AI, will drive the creation of environmentally friendly products. New sustainable energy will replace coal, petroleum, and other carbon-emitting energy sources. Innovative recycling and reuse technologies will transform waste into profit centers. For example, researchers at the University of Toronto were able to collect waste cooking oil from McDonald's and turn it into resin for 3D printing.[42]

The damage caused to the environment has been significant. Our greatest concern is that we don't know how much of it is irreversible and how soon we will be able to stop the damage so that we are not caught in an existential threat situation. However, there is great hope that future technologies and sustainable energy sources will allow us to live in harmony with our environment.

These 10 megatrends, as they become more and more pervasive and accessible, have the potential to completely transform human activities, and create new opportunities and threats for businesses at unprecedented speeds.

Exponential Platforms

All of these megatrends will be shaped and accelerated by several exponential platforms that are converging and creating powerful combinatorial change of extraordinary magnitude. We have selected eight such platforms shown in Table 2.1 to illustrate how the convergence of these platforms is driving the megatrends discussed above and producing exceptional changes to our world.

We will now turn to a discussion of each of these exponential platforms. The progress in these platforms in recent times has been astonishing, and levels of performance and capability have been realized that were pipe dreams a few decades ago. More progress continues at an exponential pace. This power can be a double-edged sword. The technology platforms themselves are morally neutral, so it is up to the innovators, developers, and other stakeholders to ensure that we harness technological prowess for the greater good.

Artificial Intelligence

Of all the technology platforms that are driving the Exponential Era, none is more prevalent and impactful than AI. As you can see in Table 2.1, this is the only technology platform that is present in every single megatrend. Its magnitude and influence cannot be overstated. For this reason, we will dedicate a longer discussion on AI than any of the other technology platforms.

Table 2.1 Convergence of exponential platforms shaping megatrends

Megatrends/platforms	AI	Networks and blockchain	Biotech	Quantum computing	IoT	3D printing	Robotics	Material science
Hyper global connectivity	✓	✓		✓	✓			✓
A longer and healthier life	✓		✓	✓	✓	✓	✓	✓
Augmented and virtual worlds	✓	✓		✓				✓
Abundance	✓		✓		✓		✓	✓
Upgraded transportation	✓	✓		✓	✓	✓	✓	✓
Instant personalization	✓	✓		✓	✓	✓	✓	
Digitization of work	✓	✓		✓		✓	✓	
Automated shopping	✓	✓			✓			
Radical interfaces	✓		✓	✓			✓	✓
Sustainable environment	✓	✓	✓	✓	✓		✓	✓

AI is already entrenched in our everyday lives and is increasingly shaping much of what we do. Consumers don't always recognize it as such, as corporate marketing experts prefer to avoid technical jargon and instead use consumer-friendly names like Siri and Alexa. But for people that are more technically inclined, the ubiquitous presence of AI is hard to miss. AI is not only displacing human activity, but also changing the very nature of companies, how they operate, and how they compete.

AI is not a new concept. In fact, its roots go back several decades. But its widespread adoption and exponential growth have only been felt in the last 10–15 years due to the convergence of three growth vectors: algorithmic advances, computing power, and data explosion.

The first vector, algorithmic advances, goes back as far as 1805 when French mathematician Adrien-Marie Legendre published the least square method of regression which provides the basis for many of today's machine-learning models. In 1965, the architecture for machine learning using artificial neural networks was first developed. Between 1986 and 1998, we saw many algorithmic advances, including backpropagation, image recognition, natural language processing, and Google's famous PageRank algorithm.

The second vector, computing power, had a significant historical landmark in 1965 when Gordon Moore recognized the exponential growth in chip power. At the time the state-of-the-art computer was capable of processing in the order of 3 million FLOPS (floating-point operations per second). By 1997, IBM's Deep Blue achieved 11 Giga FLOPs (11 billion FLOPS), which led to its victory over Garry Kasparov, the world chess champion. In 1999, the Graphics Processing Unit (GPU) was unveiled, which is a fundamental computing capability for deep learning. In 2002, we saw the advent of Amazon's Web Services (AWS) making computing power easily available and affordable through cloud computing. Computing power continues to advance today as we will see in the discussion about quantum computing.

Finally, the third vector, data explosion, started in 1991 when the World Wide Web was made available to the public. In the early 2000s we saw wide adoption of broadband, which opened the doors to many internet innovations, resulting in the debut of Facebook in 2004 and YouTube in 2005. At that time, the number of internet users worldwide surpassed 1 billion.

The year 2007 became a significant landmark. It is at this point that the technologies began converging as the mobile explosion came to life with Steve Jobs' announcement of the iPhone. By 2010, 300 million smartphones were sold, and internet traffic reached 20 exabytes (20 billion gigabytes) per month.

In 2011, a key milestone was achieved. IBM Watson defeated the two greatest Jeopardy champions, Brad Ruttner and Ken Jennings. Such achievement was made possible by IBM servers capable of processing 80 teraFLOPS

(80 trillion FLOPS). Remember that when Moore's law was pronounced in the mid-60s, the most powerful computer could only process 3 million FLOPS.

By 2012, significant progress had been made in deep learning for image recognition. Google used 16,000 processors to train a deep artificial neural network to recognize images of cats in YouTube videos without providing any information to these machines about the images. Convolutional neural networks (CNN) became capable of classifying images with a high degree of accuracy. In the meantime, the data explosion continued, with the number of mobile devices on the planet exceeding the number of humans, generating 2.5 quintillion bytes of data per day by 2017. Computing power reached new heights as Google announced its Tensor Processing Units (TPU) capable of 180 million teraFLOPS.

It is at this point in the history of AI that many people started to realize we might not be too far from achieving, or even exceeding, what is known as artificial general intelligence (AGI). To the astonishment of the world, Google's DeepMind hit another major milestone when its AlphaGo Zero algorithm learned to play on its own the games of chess, shogi, and Go. Go is a very complex game, much more challenging than chess. Not only did AlphaGo Zero learn to play by itself, but it also defeated the best computers that had been fed instructions from human experts. And it did this in only eight hours of self-play. AlphaGo Zero, by itself, came up with playing strategies that humans had never thought of before.[43]

Today, AI is capable of creating new paintings in the style of renowned artists that are indistinguishable from original paintings. The New Rembrandt, an AI creation based on thousands of scans of Rembrandt's 300 known paintings, has shocked the world due to AI's ability to maintain originality while observing the unmistakable style of a Rembrandt painting.[44] AI is also writing entire books. An AI-written novel made it to the final round in a competition for Japan's national literary prize.[45]

Remarkably, this revolutionary technology, which was first only within the reach of large corporations and governments, is available to all of us today. Most AI software is already available as open-source or as pretrained algorithms that can be accessed as a service. For example, with the introduction of the "Transformer" by Google in 2017, language understanding made significant advancements. Transformer is a deep learning architecture that was designed to increase performance on natural language tasks in an efficient way. Throughout 2019, Google, OpenAI, Microsoft, and Facebook released pretrained models of Transformer. Now anyone can download it and fine-tune it for their particular task, avoiding the huge expense of training from scratch. Hugging Face has put together all of these pretrained models under a unified application programming interface (API) with everything needed to get started.[46]

AI is transforming industries like Finance, Insurance, Retail, Entertainment, Healthcare, and more. In their book *Competing in the Age of AI*, Marco Iansiti and Karim Lakhani state that "AI is transforming the operational foundations of companies, enabling digital scale, scope, and learning, and erasing deep-seated limits that have constrained firm growth and impact for hundreds of years." They add further: "And in this new age of AI, many time-honored assumptions about strategy and leadership no longer apply." This further validates the need for a new strategic planning process — one of the many impetuses for writing this book.

In the last decade, we have seen the emergence of a new kind of company, one that is architected from the beginning to operate in a digital operating model, leveraging networks, large amounts of data, and AI algorithms. This new digital operating model is adept at functioning in much greater scale, scope, and capabilities that can overwhelm traditional companies, resulting in a level of disruption that will challenge traditional managerial methods. In order to compete in this environment, traditional firms will have to transform into a different kind of company, rearchitecting how they access and use data, react to information, and make and execute operating decisions.[47]

Networks and Blockchain

We have combined networks and blockchain as one group for easier categorization. Blockchain is a distributed technology, so it makes sense to think of it as an additional layer that allows the effective authentication and securitization of transactions that run on a network.

Networks are essentially a means of connectivity and transportation. Think of the roads and railroads that connected populations as an early version of a network that served as a catalyst for the exchange of ideas and commerce. Next, the networks became electric, in the form of power grids and the telephone. But the real exponential explosion of networks has happened in the last 50 years as the networks have gone digital. A vast infrastructure of fiber-optic cables, wireless towers, and satellites connect an estimated 5 billion people. The rest of the world's population is expected to join the conversation in the next five years.

One key technology enabling very fast connectivity is 5G. When the internet emerged, we were running 2G networks. A decade later, we jumped to 3G, and with the proliferation of smartphones, we enjoyed the much faster 4G networks that are most prevalent today. However, 4G is quickly being replaced by 5G technologies that promise to completely alter the connectivity landscape by offering much faster speeds. As covered in the hyper global connectivity megatrend discussion, a wide range of companies is investing heavily in multi-terabit constellations of satellites that will bring fast and affordable connectivity to every person in the world.

Riding on top of this global network are blockchain technologies. Many people associate blockchain with Bitcoin, or cryptocurrencies, in general. Cryptocurrencies are indeed based on blockchain technologies, but the range of usage for blockchain is much broader than just cryptocurrencies.

Blockchain contains four characteristics that are fundamental to its functioning:

- Distribution – Blockchain is distributed across the network, so no single person or entity is the custodian of the records contained in the ledgers in a blockchain. In fact, there are multiple copies of the ledger, so even if one or many are destroyed, there is a large amount of redundancy that sustains the integrity of the blockchain.
- Immutability – This enforces the validity of the information held in the blockchain by guaranteeing that if a change is made to one ledger, all ledgers must also change. It is not possible to have inconsistent information in the blockchain.
- Permissibility – This means that anyone can use it. There are no intermediaries or approvals in the process. This also means that even people considered nonbankable can now transact and participate in global commerce. But it also results in the potential rise of online black markets and other ill-intentioned uses of the technology that will require new ways to govern and police its usage.
- Transparency – Everyone on the network can see every transaction. This solves many problems, but one in particular that is very meaningful is that blockchain prevents cheating. For instance, you cannot pay someone with a blockchain digital token, and then make a copy of that token and pay someone else. Blockchain solves the "double-spending" problem that until then had been a barrier to digital currencies.

Cryptocurrencies remove the middleman. In other words, once we can transact directly with each other, there will not be a need for banks, or intercontinental money exchanges, like Western Union, that charges a fee for every transaction. Major financial firms like J.P. Morgan, Goldman Sachs, and Bank of America are very aware of this, so they are positioning themselves with their own strategies to roll out cryptocurrency-related products and services. According to Gartner, this market could exceed $3.1 trillion by 2030.[48]

Blockchain technologies are also excellent for establishing Digital IDs and validating a person's or an object's identity. For instance, a personal blockchain ID can guarantee accurate and fair voting and personal reputation scores, such as creditworthiness. It can also validate the originality of a piece of art or true ownership of a piece of land.

Remarkably, blockchain can automate the entire contracting process. Blockchain smart contracts can contain all the terms of the agreement and a set of conditions that would trigger execution. Once the conditions are met, smart contracts can automatically complete the transaction. For instance, payment can be made automatically once a shipment delivery is confirmed, eliminating many manual processes in current carrier businesses.

With blockchain, all digital objects can become smart objects. For example, an instance of a digital art object can have its own identity and be easily monetized. Blockchain will guarantee the originality of that object eliminating fake copies. Combined with AI, smart objects can become even more interesting, gaining memory and intelligence. Digital twins powered by blockchain and AI will gamify the world, creating a whole new dimension to our physical realities.

Biotechnology

Biotechnology is advancing our ability to eliminate diseases that until recently seemed incurable. Gene therapy allows the replacement of defective DNA inside a cell. Gene-editing techniques like CRISPR-Cas9 allow the repair of DNA, instead of its replacement, inside the cell. Stem cell therapies replace cells entirely. All of these techniques are revolutionizing medical treatments and bringing new hopes for a brighter future.

According to David Liu, a lead researcher at Harvard who was part of the team that unveiled CRISPR 2.0, "of more than 50,000 genetic changes currently known to be associated with disease in humans, 32,000 of those are caused by the single swap of one base pair for another." With CRISPR 2.0's precise ability to change a single letter in a single string of DNA, many of these diseases are about to be eliminated.

According to Peter Diamandis, with the advent of stem cell therapy, researchers are developing treatments for cancer, diabetes, arthritis, heart disease, macular degeneration, skeletal tissue repair, pain management, neurological diseases, auto-immune conditions, burns and other skin diseases, blindness, and much more.

As stated in the Longer and Healthier Life megatrend discussion, longevity experts believe we will soon achieve "longevity escape velocity." Scientists have identified nine causes of human decay and believe that we can address each of them through new biotechnology advances, leading to an eventual increase of more than one year in life expectancy each year. While those with greater means will adopt these developments first, the real advances for humanity will come with pervasive accessibility.

Biotechnology also promises the development of healthier foods that are more efficient to produce and environmentally sustainable, reducing greenhouse gases and consuming less water. Through stem cell enrichment we can now produce protein sources that taste great and that can feed the world without the negative effects of raising animals. Cultured meat uses 99% less land, 82–96% less water, and produces 78–96% fewer greenhouse gases.[49]

We have outlined the positive uses for biotechnologies. However, we acknowledge that, as with any other technology, biotech can be used for nefarious purposes, such as the full range of biowarfare.

Quantum Computing

Just when Moore's law appeared to have hit its end due to the physical limitations of semiconductors, quantum computing came to the rescue. When transistors approach 5 nm, they start to become unstable due to the electrons behaving unpredictably. But quantum computing offers a new paradigm where instead of the classical binary state represented by a bit, a multistate qubit, achieved via "superposition" in extremely cold temperatures provides an extraordinary level of computing power. For example, a 50-qubit computer has 16 petabytes of memory.

Quantum computing is no longer just a pipe dream. In fact, it is available to anyone who can afford it today. Rigetti Computing's Quantum Cloud Service provides virtualized programming and execution environments that can run quantum software applications. IBM Quantum provides cloud-based software access to quantum computers anytime by anyone. Quantum computing is already being used to solve real-world problems. For example, Volkswagen uses quantum computing to optimize traffic flows in congested cities.[50]

Quantum computers could spur the development of breakthroughs in science, biotechnology, AI, materials science, and much more. There are challenges of such size and complexity, that traditional computers are simply unable to handle. With quantum computing the realm of possibilities expands enormously, ushering a new wave of discoveries and breakthrough innovation.

Internet of Things

The Internet of Things, also known as IoT, refers to all the billions of devices that are connected to the internet. They include sensors, which are any electronic devices that measure things like temperature, pressure, light, or any other physical quantity and transmit the information over the network.

The number of devices on the network has grown remarkably fast. In 2009, for the first time, the number of devices surpassed the number of people on the network. By 2015, that number more than tripled, reaching 15 billion. By 2030, 125 billion devices are expected to be on the internet.[51]

These sensors will be monitoring and sensing all facets of our environment, nonstop, creating a global sensor matrix that will allow us to know almost anything, anytime, anywhere. And the benefits are manifold.

For example, farmers will have the ability to know moisture content in the soil and the air, improving the effective utilization of water. Weather forecasters will have many more data points, including sensors placed in the middle of the ocean, that will help develop more accurate and precise forecasts. Smart appliances connected to the network will create more convenience and efficiencies. Wearable technologies will allow us to monitor our health and provide vital information to our health providers.

But the proliferation of IoT devices also raises a new level of privacy concerns. Surveillance cameras everywhere with AI-based face recognition capabilities will not only enhance security, but also track every person's moves. Personal assistant devices in homes raise the question of whether they are always listening and who has access to these conversations. Like any other technology, IoT can have both positive and negative repercussions.

3D Printing

3D printing is a fabrication technique for building three-dimensional structures based on the digital characterization of objects. The fabrication is achieved by adding layers of material until the final object is created, a technique called additive manufacturing.

The first 3D printers were made available in the 80s, but back then they were hard to program, difficult to use, and could only work with plastic. Today 3D printers are highly efficient and easy to use. They can use hundreds of different materials such as metals, rubber, plastic, glass, concrete, and even organic materials such as cells. This allows 3D printing to fabricate a wide variety of complex devices from jet engines to apartment buildings, to circuit boards.

3D printing can use different techniques for fabrication, including extrusion, by depositing layers of melted plastic filaments; resin, a photosensitive liquid that is solidified via a source of light; and powder, where powdered material is sintered or melted and the grains of powder are bonded to obtain a solid structure. With this level of flexibility, GE was able to reduce the number of parts in its advanced turboprop from 855 to just 12 and achieve a 100-pound weight reduction with a 20% improvement in fuel burn.[52]

3D Printing can also be used in construction to build housing that is eco-friendly, affordable, scalable, efficient, and that allows flexible design. A Chinese company combined 3D printing with modular construction to erect a 57-story building in 19 days.[53] In 2019, California's Mighty Building was able to meet US building code standards to produce 3D printed single-family homes that cost a third of traditional houses.

Even more exciting is bioprinting. This is a manufacturing process where biomaterials such as hydrogels are combined with cells and growth factors to create tissue-like structures. It uses a material known as bioink to build structures layer by layer. A company called Prellis Biologics is printing capillaries, and IVIVA Medical is able to produce kidneys.[54] 3D printed organs are predicted to be available in the market in 2030.

3D printing technology is disrupting supply chains, manufacturing, transportation, and warehouses. This single exponential platform is transforming and rearranging the ecosystems of the entire $12 trillion manufacturing industry.

Robotics

Robots are increasingly becoming ingrained in all aspects of life. Factories across the globe use robots to manufacture all kinds of products, from automotive to electronics. They are also being used in warehouses to sort, pick, and stack units. Today, the use of robots is focused on repetitive jobs or those that are dangerous or dirty. Tomorrow, robots will be used in a much wider range of jobs from surgery to everyday maintenance and customer service.

Domino's pioneered the first pizza delivery robot, DOM.[55] Then Starship Technologies created general-purpose robots designed to leverage several technologies such as cameras, GPS, and AI-driven natural language processing to automate home deliveries.[56] This suggests to us that the days of the UPS delivery person are numbered.

One type of robot that is being used effectively in a variety of use cases is drones. From their early days as just toys to today's usage in life-threatening situations, drones have come a long way. Drones have delivered medical supplies in relief efforts for Hurricane Sandy, Typhoon Haiyan, and many other natural disasters like floods and earthquakes.

In 2016, Amazon announced Prime Air, a drone delivery system that promises delivery in 30 minutes or less. Immediately, several other companies followed suit, including 7-Eleven, Walmart, Google, and Alibaba. Although there are many skeptics, and regulations to work through, the head of the Federal Aviation Administration's (FAA) drone integration department says that these types of deliveries are closer than the skeptics think, given the number of applications the FAA is working to approve.[57]

Intelligent robots, powered by AI, will likely replace most manual labor in factories, warehouses, stores, construction, agriculture, and many other industries. While the end of repetitive and dangerous work improves the human condition, it does raise concerns about the re-education of the labor force and the implications for those that are most vulnerable to these potential changes to the labor market.

Material Science

Materials are the raw ingredients that go into products. Materials are easily taken for granted because most people don't have a good grasp of the intricate work that goes into creating the materials that are used in today's products. Behind the scenes, material science has enabled the creation and affordability of the everyday items we consume today. As an example, Omkaram Nalamasu, the Chief Technology Officer at Applied Materials, explains that "if you built a version of today's smartphone back in 1980, it would cost something like $110 million, be fourteen meters tall, and require about two hundred kilowatts of energy. . .that is the power of advances in materials [science]."

Material science is bringing huge improvements to the solar industry. One of the newest materials that promises to increase the conversion of sunlight into electricity from about 15% using today's technology to a theoretical limit of about 65% is perovskite, a light-sensitive crystal with the potential to make solar energy affordable to anyone.[58]

But perhaps the most exciting advances in materials science is being made at the atomic level with nanotechnology. Nanotechnology promises to take basic materials like water and air, atom by atom, and reconstruct them into just about anything. For example, scientists can take carbon dioxide from the atmosphere and turn it into very strong carbon nanofibers for use in manufacturing.

Another example is nanographene. Graphene is a honeycomb sheet of carbon atoms. With nanotechnology, it can be added to aluminum and magnesium to make them stronger than titanium, it can turn plastics into electric conductors, and it can create resins that are corrosion resistant. Nanographene also promises to be effective in cleaning up pollutants.

Nanomaterials are driving the creation of lighter and more affordable products like airplanes, automobiles, bicycles, power tools, and much more. The applications are practically endless.

As we have seen throughout this chapter, the convergence of exponential platforms is creating new ecosystems and driving megatrends that will have a profound impact on every aspect of human life. This represents an incredible opportunity for businesses that are agile and innovative. It also creates many challenges to existing business models. In the next chapter, we will examine several businesses that are managing these changes well and those that are struggling to adapt to these new times of accelerating changes and disruptive forces.

CHAPTER 3

Animals of the Exponential Kingdom

We have seen how exponential platforms are shaping new ecosystems and creating megatrends that will have a profound impact on life and business. For some businesses this represents an opportunity, for others, it is a threat. We have argued that in order to compete in this environment, traditional firms will have to transform into different kinds of companies, rearchitecting their operating mechanisms and developing digital business models.

In this chapter, we will review various companies that we have playfully classified as different types of animals with unique attributes in this wild exponential kingdom: Flash Boiled Frogs, Disruptive Unicorns, Fast and Furious Gazelles, Dancing Elephants, and Dominating Gorillas. We have selected two companies in each category to represent their species.

Flash Boiled Frogs

The boiling frog is a fable describing a frog being slowly boiled alive. The premise is that if a frog is put suddenly into boiling water, it will jump out, but if the frog is put in tepid water which is then brought to a boil slowly, it will not perceive the danger and will grow too weak to jump and slowly be cooked to death.

Many companies behave like boiling frogs. Some completely miss the fact that changes on the horizon – technological, societal, competitors, markets, or other business factors – represent a real threat to their business; they just

The Exponential Era: Strategies to Stay Ahead of the Curve in an Era of Chaotic Changes and Disruptive Forces, First Edition. David Espindola and Michael W. Wright.
© 2021 by The Institute of Electrical and Electronics Engineers, Inc.
Published 2021 by John Wiley & Sons, Inc.

don't see the eventual magnitude, speed, and direction of these changes early enough to respond. Others see them but don't intercept them, meaning, just like the boiling frog, they can sense that their environment is changing, but they don't realize the gravity of the threat and don't react until it is too late.

In today's Exponential Era, the changes are accelerating beyond the slow boil of old, and companies that don't react become what we call "Flash Boiled Frogs." At the rate of change they are observing in the moment, by the time they discern what is happening, the condition has moved from an opportunity to a threat. The slowness of companies' responses in the face of exponential change renders them boiled in a flash.

There are many examples of companies that have failed to see or react to the changes around them, some of which we will mention later in the book. For this chapter, we have chosen two classic cases of monumental declines that exemplify the dangers of missing business inflection points.

Nokia

In November 2007, the cover of Forbes magazine bravely pronounced: "Nokia, one billion customers – can anyone catch the cell phone king?" At the time Nokia was earning more than 50% of all the profits in the mobile phone business and was by far the most recognized brand in the industry. It was hard to imagine back then that any company could ever overcome such prominent dominance.

That same year Steve Jobs introduced the iPhone. And, as they say, the rest is history. The rhetorical question in the cover of Forbes would prophetically foretell the destiny of what was then the most recognizable brand in the industry. The implausible change of fate for Nokia culminated in a sale to Microsoft in 2013. From there, the Nokia brand practically evaporated.

How could Nokia have missed the smartphone inflection point? This was a highly adaptive company that moved from rubber galoshes to paper, to phones. It spent enormous amounts of money on R&D and was considered a technological powerhouse. Ironically, it introduced the first smartphone as far back as 1996 and built an internet-enabled touchscreen prototype in the late nineties. In that same year, Nokia declared the cellphone as the eventual "remote control for life."

So, what went wrong? In August 2011, Marc Andreessen wrote the famous Wall Street Journal article titled *Why Software Is Eating the World.*[59] Nokia was a hardware company with great expertise and pride in building physical devices, but where software engineers were marginalized. What they didn't understand was that the smartphone business was all about the software. Nokia missed the "software-eats-the-world" inflection point.

It took more than a decade, from the time smartphones were prototyped to when Nokia in its original form ceased to exist. Back then, this was fast. Today, we have seen entire industries, like taxis and newspapers, be disrupted in a fraction of that time. The future will be even faster.

Blockbuster

Around the same time that Nokia entered its downfall, another fascinating frog tale was developing. John Antioco – a retail genius with a long history of success – was the CEO of Blockbuster, a video rental powerhouse with thousands of retail locations, millions of customers, massive marketing budgets, and superefficient operations. Blockbuster was the category king of the video business, with absolute dominance over its competitors.

In the year 2000, John Antioco was approached by Reed Hastings – the founder of a fledgling company called Netflix – who proposed a partnership where Netflix would run Blockbuster's brand online and Antioco's firm would promote Netflix in its stores. Hastings got laughed out of the room.

While Antioco basked in the sun enjoying Blockbuster's success, Hastings kept growing his small mail-order video business. At the time, Netflix did not seem to represent a threat. It was hard for people to find it because it did not have any retail stores. And since customers received their videos by mail, they couldn't just pick up a movie for the night on their way home.

Still, customers loved the service and told their friends, and slowly but surely, Netflix kept growing. By 2008, Netflix had taken some of Blockbuster's market share, but Blockbuster refused to acknowledge the threat. Just like the boiling frog, it certainly sensed that the water was getting hotter, but even with a change in CEOs, it remained in denial.

"Neither Redbox nor Netflix are even on the radar screen in terms of competition," said Jim Keyes, CEO of Blockbuster, in December 2008. By 2010, Blockbuster was bankrupt. Looking back, the level of defiance and arrogance that defined the Blockbuster story just seems so surreal. This can serve as a wakeup call to companies that find themselves in denial today.

Blockbuster's business model was based on a retail mindset that believed a strong physical presence represented a barrier to entry and a competitive advantage. They simply could not see how a mail-order business could be a threat. What they missed is that Reed Hastings had much bigger plans. The inflection point did not become apparent until it was too late. Netflix did not become an undeniable threat until they adjusted their business model to leverage the ability to stream videos via the internet.

Blockbuster completely missed the network inflection point, and in less than a decade, went from king of the hill to a Flash Boiled Frog, experiencing a sharp drop in revenues in the last two years before bankruptcy.[60]

While frogs were being boiled, other animals were appearing on the scene.

Disruptive Unicorns

Unicorns are private companies with a valuation of over one billion dollars. We refer to them as "Disruptive Unicorns" due to their explosive nature. As of this writing, there are more than 400 of them around the world.[61] They tend to be the new generation of companies that grow from startup to global enterprise at incredible speeds. They typically adopt a digital business model that takes advantage of the convergence of exponential platforms described in Chapter 2.

Here we present two unicorns that are disrupting industries and growing spectacularly. We have carefully chosen 2 out of more than 400 candidates to illustrate the power of digital business models. One has become a household name. The other you probably never heard of, despite their 75-billion-dollar valuation.

Airbnb

Airbnb is the largest hotel chain in the world, yet they don't own a single hotel room. That is the power of digitization. They simply leverage the assets – in this case, available spare bedrooms, from the crowd – and run a digital operating model that gives them scale, efficiency, and agility. With a valuation of 31 billion dollars and with plans to go public in 2020, Airbnb has disrupted the hospitality industry, colliding with incumbent behemoths.

Airbnb's implausible rise to dominance in its market is one of the most fascinating stories coming out of Silicon Valley. In 2007, roommates Brian Chesky and Joe Gebbia could not afford to pay rent in pricey San Francisco. In an attempt to raise a few extra dollars, they came up with a "crazy" idea. They knew there was a conference coming to town and were aware of the difficulty in finding hotel availability during the conference, so they decided to turn their loft into a bed-and-breakfast of sorts and offered air mattress sleeping arrangements for $80 per night. Two men and a woman took them up on the offer.

Realizing this could turn into a real business, the roommates invited another old roommate, Nathan Blecharczyk, to join the startup. In 2008, after a couple of failed launches, they sought funding from angel investors but were rejected by all the 15 introductions they were able to arrange. Broke and in debt, they pushed their creative abilities to the next level. For the 2008 Democratic National Convention, they transformed cereal boxes into Obama O's and Cap'n McCains and sold them on the street for $40. Each box had a limited-edition number and some information about the company. The ingenious marketing stunt resulted in $30,000 in desperately needed funding.

This was enough to catch the attention of Paul Graham from Y Combinator, one of the most prestigious incubators in Silicon Valley. In 2009, the team spent three months at the incubator perfecting their product. Still, most investors could not wrap their heads around air mattresses in a living room being a viable business. The founders' background in design was fundamental in helping them fine-tune their offerings. They personally spent time with their hosts, took pictures of their spaces, and wrote stories about them. Finally, in 2009, the company was able to raise $600,000 from Sequoia Capital. This was the start of their phenomenal rise to prominence. Chesky famously lived exclusively in Airbnbs to learn everything he could about the business. By 2011, Airbnb was in 89 countries and hit 1 million nights booked on the platform, leading to a $112 million investment by big VCs at a valuation of $1 billion.[62]

From its humble beginnings, Airbnb has grown to become a real threat to traditional hospitality firms. While hotel chains like Marriott and Hilton require tens of thousands of employees to manage thousands of properties, Airbnb focuses on aggregating data and developing algorithms that track and manage community and property owners. Airbnb has been able to scale its operations based on an AI foundation while keeping a small organization. In less than a decade, it has been able to offer more than 4.5 million rooms, three times as many as Marriott was able to accumulate over 100 years, and yet it employs only 12,736 people compared to Marriott's 177,000.

This digital operating model provides Airbnb with many advantages. The network effects permit Airbnb to quickly scale, learn, and respond very quickly to customer needs. It uses its vast data sources to acquire new customers, identify their needs, and optimize their experiences. As it accumulates more data, Airbnb is able to leverage machine learning to analyze risks and gain new insights through experimentation. It can try a broad variety of new offerings such as entertainment and travel-related services without spending a huge amount of money or incurring unnecessary risks. This improves their ability to learn quickly and drives opportunities for additional value creation and capture.

In the meantime, Marriott is struggling to operationalize mergers and rethink its operating model against the threat of data-driven agile competition that surfaced suddenly, disrupting its business.[63] Obviously, we don't know what will happen to the hospitality industry post COVID-19. Regardless of the outcome, Airbnb will stand as an example of a Disruptive Unicorn.

ByteDance

If you have never heard of ByteDance, you may be surprised to learn that in 2018 it was the world's most valuable startup, at a 75-billion-dollar valuation, surpassing Uber, Airbnb, and SpaceX. Founded in 2012 in China by the then 27-year-old entrepreneur Yiming Zhang from an apartment bedroom, it has taken the world by storm with its viral social media app TikTok. Rebranded from Musical.ly and Douyin, TikTok has accumulated over half a billion monthly users worldwide. The app has been downloaded nearly 80 million times in the United States.

One of its first products is a hugely popular news aggregation service known in China as Jinri Toutiao (which means "Today's Headlines"). Toutiao is powered by AI to send personalized stories to its more than 100 million daily users based on their personal preferences. Their spokesperson explains how the company leverages AI to power its apps: "Artificial Intelligence powers all of ByteDance's content platforms. We build intelligent machines that are capable of understanding and analyzing text, images, and videos using natural language processing and computer vision technology. This enables us to serve users with the content that they find most interesting and empower creators to share moments that matter in everyday life to a global audience."[64]

The company is expanding rapidly in several areas to grow revenues. In March 2019, its subsidiary Lark Technologies launched a product called Lark which provides collaboration capabilities.[65] In May 2019, TechCrunch reported that ByteDance was exploring music streaming service.[66] The same month, the Financial Times reported that ByteDance was looking to launch its own smartphone preloaded with apps that the company makes.[67] The company also appears to be entering the education market by acquiring the patents and team of Smartisan. A spokesperson spoke to CNBC about the acquisition: "The goal of this acquisition is to incorporate the team's strength to build upon ByteDance's initiatives in the education space, specifically for upcoming education hardware."

Flushed with cash from top VCs like Softbank and Sequoia Capital, ByteDance is investing aggressively to tap into the lucrative western social media market and many other areas of opportunity. Adopting a digital operating model powered by AI, ByteDance is vying to become a global news provider that employs no editors and no reporters. In a 2017 interview, Zhang confirmed this sentiment, adding further: "The most important thing is that we are not a news business. We are doing very innovative work. We are not a copycat of a US company, both in product and technology."[68]

Just like Airbnb and ByteDance, hundreds of unicorns are creating digital business models that are disrupting every industry. Powered by AI, these companies are able to operate with a much smaller number of employees, and yet scale at a much faster pace than their incumbent competitors.

We now turn to another fast-moving species in the exponential kingdom.

Fast and Furious Gazelles

Researcher and author David Birch coined the term "gazelle" in his book *Job Creation in America: How Our Smallest Companies Put the Most People to Work*, to represent fast-growing companies. According to his definition, gazelles – at a minimum – double their sales every four years, from a base of at least $100,000 in revenues.[69] We like to call them "Fast and Furious Gazelles." For the purposes of this book, we will distinguish Fast and Furious Gazelles from Disruptive Unicorns by excluding private companies with a one-billion-dollar valuation or more from the group of gazelles.

We have selected two companies that meet the definition above to represent the species. They are both very familiar to you and their amazing stories and aggressive ambition to dominate their respective industries through the use of exponential platforms are perfect illustrations of the types of companies that understand how to take advantage of opportunities in the Exponential Era. What is intriguing about both companies is that they have made their fair share of mistakes, but that has not deterred them from their quest to thrive in this fast-changing environment.

Uber

The Uber story begins when two friends, Travis Kalanick and Garrett Camp, who had previous successful exits as entrepreneurs, met at the LeWeb conference in Paris in 2008. Allegedly, the friends had difficulty finding a cab on a winter night during the conference and had the idea for a timeshare limo service that could be ordered through an app.

When Camp returned to San Francisco, he bought the domain name UberCab.com and began working on a prototype in 2009. In 2010, they tested the idea in New York using only three cars, then launched in San Francisco, leading to an initial round of $1.25 million investment. A cease-and-desist letter involving the use of the word "cab" resulted in the company shortening its name to just Uber. A series of funding rounds allowed the company to expand its services to multiple cities in 2011. In 2016, the company received an astounding $3.6 billion in funding from Saudi Arabia.

Despite fierce opposition from the taxi industry and competition from other ridesharing firms like Lyft, the company kept growing and expanding its services. A series of scandals and missteps by the leadership team led to the ousting of Kalanick as CEO in 2017, when Dara Khosrowshahi, former CEO of Expedia, took the reins.[70]

Delivering more than 15 million rides a day, Uber has transformed the personal transportation landscape. And this is just the beginning. Uber is a key player in the autonomous vehicle movement. According to Jeff Holder, who heads up its AI lab and autonomous car group, the transition to autonomous vehicles will happen faster than anyone expects: "Already over 10 percent of millennials have opted for ridesharing over car ownership. But this is just the beginning. Autonomous cars will be four to five times cheaper – this makes owning a car not only unnecessary but also expensive." But it is not just vehicles driving themselves on the road. Uber has much bigger ambitions, as Holden explained from the stage at Uber's Elevate conference: "Uber's goal is to demonstrate flying car capabilities in 2020 and have aerial ridesharing fully operational in Dallas and LA by 2023. Ultimately, we want to make it economically irrational to own and use a car."[71]

What Uber and other ridesharing companies have been able to do is build a datafication layer around transportation. They built an infrastructure that generates data about individual transportation preferences, supply and demand, and the flow of traffic, at unprecedented levels. This is made possible by a network effect: the value created by Uber attracts more riders, and the more riders there are, the more drivers are attracted to use the service, creating a virtuous loop.

Uber's network has allowed it to expand its offerings, connecting, for instance, food providers through Uber Eats, healthcare providers via Uber Health, and expanding into transportation with Uber Freight. Like other companies that thrive in the Exponential Era, Uber aggressively pursues experiments and is not afraid of failures. For example, Uber Freight threatens to disrupt the multibillion-dollar logistics industry by connecting freight carriers' supply and demand. Uber Freight Plus provides carriers with discounts on tires, fuel, mobile phone plans, and more.

In addition, the company is exploring opportunities with UberPool, offering ridesharing across multiple users, and Cargo, connecting its rider network with retailers to sell products to riders while they are a captive audience. Uber has even experimented with UberKITTENS – where users pay to cuddle with kittens – and Uber ice cream delivery. The combination of experiments with the accumulation and analysis of an extensive amount of data allows the company to accelerate its learning and continuously optimize the value it creates.

Uber's digital operating model, data-driven AI-based decision making, large quantities of capital, experiment-driven mindset, and network effect advantages, places the company in a league of its own where moving fast and disrupting existing markets is the name of the game.

Netflix

Netflix is another household name Fast and Furious Gazelle with a fascinating story, a few stumbles, and the eventual dominance of its market. Started in 1997 by serial entrepreneurs Marc Randolph and Reed Hastings, in its humble beginnings the company was just a website-based movie rental service that allowed people to rent DVDs online and receive the ordered movies via the postal service. After watching the movie, users would simply mail the DVDs back – the users could keep the DVD as long as they wanted but could only order a new one after returning the previous one. This would prevent the user from having to pay the infamous late fees that Blockbuster was notorious for. Allegedly, Reed Hastings has attributed his experience of being charged a $40 late fee as the inspiration for starting the company, but his co-founder has called this story a "convenient fiction" used as a marketing ploy.[72]

Reed supplied the initial $2.5 million seed capital to start the company which launched its services in 1998 with just a few employees and less than 1,000 titles at the time. In 1999, it introduced a subscription service offering unlimited DVD rentals for one monthly low price. In 2002, the company went public, and by 2006 had 4.2 million users. In 2007, it introduced streaming, and from there it expanded its services worldwide.[73] Today the company is a dominant media player.

Throughout its fast-rising journey, the company had a few challenging and defining moments. As related earlier in the chapter, Hastings approached Blockbuster and allegedly tried to strike some sort of partnership, and even offered Netflix for $50 million, which Blockbuster readily dismissed. Apparently when Netflix made the offer to Blockbuster it was hemorrhaging money and feared an outright takeover that could potentially destroy its brand. Luckily, it got rejected and was able to survive. Then in 2011, when streaming was becoming a key offering that would drive the future of the company, it made the fateful mistake of separating the DVD business into a subsidiary called Qwikster. Instead of paying a single $9.99 subscription price for both streaming and DVDs, customers would have to pay separate fees that combined totaled $15.98 per month. Customers revolted as they perceived this to be an unjustifiable price increase, resulting in some 800,000 defections in what appeared to be the beginning of the end for Netflix.[74]

Despite the challenges, Netflix not only survived but thrived, gaining a 51% share of all streaming subscriptions, earning $4.5 billion in annual revenues, and reaching a market cap of $150 billion. Fundamental to this achievement was the adoption of digital technologies; first, with streaming, which allowed it to scale worldwide at a pace that would not be possible with a mail-order business. Hastings foresaw the advent of streaming as a dominant technology, but he did not always get the timing right: "In 1997, we said that 50% of the business would be from streaming by 2002. It was zero. In 2002, we said that 50% of the business would be from streaming by 2007. It was zero. . . Now streaming has exploded. . . We were waiting for all these years. Then we were in the right place at the right time."[75] In our view,

Netflix is a classic example of a company being alert, prepared, and patient enough to intercept the horizon.

But streaming was only part of the equation. Netflix has completely transformed the media industry by personalizing the user experience through the power of AI. From the early DVD days, the company was already developing its recommendation engine which suggested movies based on users viewing history, ratings, and other users' preferences. With streaming, Netflix was able to track much more, such as when users pause or skip during a show. They could even personalize the movie thumbnail that is shown to viewers. Netflix also uses data and AI algorithms to decide which content to create on its own – that's how they evaluated the potential for *House of Cards*. Cindy Holland, Vice President of Original Content noted in an interview with Iansiti and Lakhani: "We have projection models that help us understand, for a given idea or area, how large we think an audience size might be, given certain attributes about it. We have a construct for genres that basically gives us areas where we have a bunch of programs and others that are areas of opportunity."

Just like Uber, Netflix relies on its AI-based experimentation platform to validate hypotheses, running thousands of small experiments and learning throughout the process. Every significant product change goes through A/B testing before it is made part of the product. A/B testing is a way to digitally test different versions of a product or service to see which one is favored by users. This fully automated experimentation platform allows the company to run experiments at scale and enforce scientific rigor to their decisions. The benefits are unquestionable. Netflix's digital operating model has allowed it to scale to 150 million subscribers in 190 countries, consuming 15% of all the bandwidth of the internet – a truly remarkable feat.

Dancing Elephants

Having visited with the Fast and Furious Gazelles, we now turn our attention to the "Dancing Elephants." These are companies that despite their enormous size, are able to move from one "S-curve" to another, navigating multiple inflection points. Granted, they don't always dance as gracefully as their shareholders would like to see, but they have relentlessly shown the ability to not only survive but thrive in new areas of business opportunities. The quintessential dancing elephant was presented to you in the Introduction: IBM. IBM is perhaps the best representative of the species, and was crowned as such in Lou Gerstner's classic *Who Says Elephants Can't Dance?* The other Dancing Elephant that we have chosen to discuss in this section has also been mentioned in other parts of the book: Microsoft.

IBM

What is most remarkable about IBM is how it has been able to transform itself several times throughout its long history of over 100 years. We showed in Figure I.2 in the Introduction how IBM moved through several S-curves, starting with counting machines and typewriters, then moving into mainframe computing, then client-server software, then consulting and services, and eventually to the cloud.

The business world was surprised by the selection of Lou Gerstner, then CEO of RJR Nabisco to take the helm as CEO of IBM in the early 1990s. Gerstner had no technical background, and his experience was mostly in selling to consumers, not in the business-to-business world that IBM operates in. At the time, IBM was struggling, and its share price dropped from $43 in 1987 to $13 in the 1990s. Many experts predicted the demise of the company within a few years.

What the board was looking for was someone who could change the company's culture – someone who could motivate people, build relationships, and change behaviors. They believed the kind of strategic and cultural change that Gerstner had achieved at American Express and Nabisco was just what IBM needed.

The first thing Gerstner did when he started at IBM was to spend time with people. He visited with customers and talked to frontline employees to get a feel for what was going on. He performed "deep dive" sessions where anyone who had relevant information, regardless of rank, was invited. This helped him understand the kind of challenges he was up against and build trust within the organization, a key component of his effort to influence and change the company.

Gerstner's key insight, which could only be obtained by spending endless hours getting information directly from the source, was that customers preferred to buy from vendors that could provide an integrated solution at a lower cost. Many smaller companies were not in a position to offer integrated solutions, but IBM, given its size, broad capabilities, talented human resources, and vast geographical presence, was in a great position to offer customers such integrated solutions and service those solutions better than competitors. Consequently, Gerstner decided to abandon his predecessor's plans to break up the company into independent units and kept the company together. He also listened to customers and lowered costs for both hardware and software.

Gerstner also understood that his executives would have to change their behavior, so he made several changes to that effect. For instance, he encouraged executives to interact with customers at every possible opportunity. He created a customer-centric culture to change the perception that IBM was not responsive and hard to do business with. He emphasized teamwork and highlighted the importance of hands-on management. He encouraged more risk-taking and removed fear from the organization, encouraging open and candid communications. He also encouraged promotion from within to grow the bench and improve the leadership team.

These are just a few of the changes that Gerstner implemented. He made several other key decisions, like selling companies that were not a strategic fit, and combining services and software around a network model. But most importantly, he changed IBM's culture, taking the company to a new level of prominence and credibility in its industry.[76]

IBM's future continues to unfold as the company continues to evolve. Most recently it faltered again under the leadership of Ginni Rometty, resulting in the announcement that Arvind Krishna, the head of IBM's cloud division, will take on the role of CEO. We don't know if Krishna will be able to orchestrate another turnaround as Gerstner did, but the one thing Gerstner has shown us is that, under the right leadership, the elephant can still dance.

Microsoft

The other Dancing Elephant we will discuss is Microsoft, which does not have as long a history as IBM, but it too has gone through an incredible journey, jumping from king of software to a precipitous fall, and then back again to prominence. Microsoft missed the most important software inflection points of the early twenty-first century, including mobile, social, and cloud. In 2007, Paul Graham, the influential Silicon Valley co-founder of Y Combinator published a post titled "Microsoft is Dead."[77]

In 2014, Satya Nadella was named CEO, replacing the long-time Microsoft leader Steve Ballmer. In just five years, Nadella's deep changes within Microsoft resulted in it rising to become one of the top technology companies in the world. For a moment in 2018, it became the world's most valuable publicly traded company.

The changes introduced by Nadella are manifold. On the technology side, a concerted effort to turn Microsoft into a cloud company paid many dividends. The company also changed its attitude toward its rivals and opened up to new partnerships, expanding its market reach. But the most profound and impactful change that Nadella introduced was changing Microsoft's culture.

Known for its internally hypercompetitive and even hostile environment, Nadella had his work cut out for himself in his relentless pursuit of a more gentle, cooperative work environment. Nadella's chosen theme for a senior leadership meeting was "empathy," something leaders at Microsoft were not accustomed to. In his book, *Hit Refresh*, Nadella states that "Empathy grounds me and centers me." As Nadella looked at the transformation that he wanted to lead at Microsoft, he realized he needed to change the mindset of the leadership team and the employees. He found that he needed to change the company's culture from what researcher Carol Dweck calls a "fixed mind-set" – which was manifested at Microsoft as employees needing to be right – to a "growth mindset," where the focus is on learning and being open-minded to new information.[78]

If there is one thing that we can learn from the Dancing Elephants, it is that transforming a large organization is not just about technology – it's about mindset and culture. This is the most important message that we can leave with you after studying how both IBM and Microsoft struggled but eventually found their way to prominence again.

Dominating Gorillas

We now come to the final species we will examine in this chapter, the "Dominating Gorillas." These are companies that are so large and powerful that they impact many aspects of our lives, from the day-to-day products and services we consume to our privacy and even politics. If you live in a modern society, it is hard not to be affected by these behemoths. We have chosen Google (and its parent company Alphabet) and Amazon to represent the species.

Google

The Google story starts in the mid-90s with two PhD students at Stanford University, Larry Page and Sergey Brin, who were working on a search engine called BackRub. Their key insight was that backlinks, which are links from websites pointing back to one's own website, is an indicator of the website's significance and authority. Allegedly, in 1998, Andy Bechtolsheim wrote a check for $100,000 as Google's first investor, but it took two weeks to cash the check because they didn't have a legal entity set up yet. Google eventually incorporated in September 1998 and opened its first office in the garage of a friend that Page and Brin sublet. In an amazing stroke of luck, they tried to sell the company for one million dollars but found no buyers. Yahoo turned them down, not just in this one instance, but again in 2002. From there, Google just kept growing with additional infusions of venture capital, hiring thousands of employees, and making several acquisitions. Another Silicon Valley legend was born.[79]

Today Google dominates the advertising industry as well as the mobile operating system market and continues to explore many other areas of expo-nential opportunities, including AI, virtual reality, autonomous vehicles, satellite communications, and even healthcare. Google's search engine processes 3.5 billion queries per day. In 2017, Google reached $95 billion in ad campaign revenues. Together with Facebook, they capture roughly 25% of all global advertising expenditure.

Google is not afraid to explore. As discussed in the Introduction, they are proud to disclose a list of 190 products and services that they have shut down. Powered by an infusion of cash from its advertising business, and a moonshot mindset, Google is today one of the most influential companies in the world.

What is most remarkable is not what Google has achieved so far, but where they are going. Google is leading the race on AI research. It stunned the world in 2017 when it demonstrated the power of AlphaGo Zero which became the undisputed best Go player in the world in just 40 days, learning the game all by itself. Google continues to invest heavily in developing AI that is capable of building even more powerful AIs, accelerating technological advancements to unprecedented levels.

Google is obtaining a stunning amount of data that will strengthen its AI position even further. Every query on its search engine, every email on Gmail, and every Google assistant request strengthen Google's understanding of our most intimate and private information while training its AI arsenal. Google is also racing to gather healthcare information through Verily Life Sciences, Alphabet's healthcare division, which is developing a large range of internal and external sensors that monitor everything from blood sugar to blood pressure, blood chemistry, and more.

Google is even trying to solve the one problem that has defeated all humans throughout history: death. The Atlantic caused a stir in 2013 with the headline "Google Wants to Cheat Death."[80] Google is putting its size, money, and brilliant minds behind its antiaging efforts, which could be considered one of its most ambitious projects. If Kurzweil is right, longevity escape velocity may be just around the corner, and Google will do everything in its power to accelerate what could be one of humanity's greatest achievement yet.

It is hard to imagine a company demonstrating more agility and willingness to experiment at scale and that is better positioned to lead the future than Google. Except for one.

Amazon

We could not think of a more ideal company to represent the Dominating Gorillas than Amazon. Jeff Bezos started Amazon in 1994. Inspired by the growth of the internet and applying what he refers to as a "regret minimization framework," Bezos left his cushy Wall Street job and headed to Seattle to follow his dreams. The initial funding came from his parents, but eventually, he raised $8 million from Kleiner Perkins Caufield in 1995. Since then the company has kept on reinvesting in the business, growing both organically and through dozens of acquisitions.

The "Everything Store" leapfrogged the king of big-box stores, Walmart, reaching the amazing market cap of one trillion dollars. Given Bezos's ambition and incredible ability to execute, we may soon be able to strike the word "Store" from its moniker and just call it the "Everything."

Amazon is investing heavily in AI capabilities and is acquiring data about user habits and preferences at an astounding rate. Amazon Echo was a brilliant move that allowed Amazon to get into people's homes and remove friction from shopping. On average, Amazon Echo users spend more than Amazon prime subscribers.[81] The company is at the forefront of leading-edge technologies that will completely transform the shopping experience. For example, in 2017, it acquired the body scanning startup Body Labs.[82] Having your exact measurements and a trove of data about your preferences, Amazon's AI will take personalized shopping of custom clothing to a whole new level.

In 2012, Amazon acquired Kiva Systems which allowed it to deploy 45,000 robots throughout its fulfillment centers, giving it the ability to process 306 items per second.[83] In addition to automating its warehouses, Amazon is bringing consumers the ability to shop at an automated brick-and-mortar store with no human interaction. Amazon Go allows customers to walk in, pull items from the shelves, and walk out, being automatically charged for the items purchased.

What Amazon has learned in building the automation systems it used internally has also turned into new sources of revenue. Amazon Web Services (AWS) has made compute and storage capabilities available to the masses very cost-effective. It now also offers SageMaker, which allows customers to gain data insights by using Amazon's pretrained algorithms and tools – think of it as AI-as-a-Service.

And it doesn't stop there. Amazon is becoming one of the top streaming platforms. It is also spending $5 billion on original content. Furthermore, it is leading in the area of logistics and transportation, investing in terrestrial and aerial automated vehicles. In 2019, Amazon announced project Kuiper, a constellation of 3,236 satellites designed to provide high-speed broadband connectivity to the entire world.[84]

Amazon's ability to scale and extend its scope is mindboggling. It is able to do this because it harvests the advantage of digital technologies that can scale easily regardless of size and complexity. Bezos built the entire company on top of a highly efficient and modular software architecture that permits small teams to work independently but effectively. He determined that teams should be no bigger than the number of people that can be fed by two large pizzas. Iansiti and Lakhani explain that small teams at Amazon "can work independently while respecting clear architectural rules that enable teams to share common code and aggregate data across applications."

The Dominating Gorillas understand that in order to continue scaling and broadening their scope they need to leverage AI running on top of modular software architecture and large amounts of aggregate data, driving the organization to continuously learn through ongoing experimentation.

Throughout the chapter we have seen companies that have done exceedingly well, taking advantage of the opportunities offered by the Exponential Era, and companies that have not done so well, unable to keep up with the fast changes that characterize this era. In the next chapter, we offer an approach to strategic planning that we believe will help all companies do exceedingly well in the Exponential Era.

SECTION TWO

CHAPTER 4

Introducing SPX: Strategic Planning for the Exponential Era

S
o far in the book, we have provided the context for the realities of the Exponential Era. We have contended that the Exponential Era is unlike any other era we have seen before. We have made the case that the speed of change, the confluence of technological advancements, and the creation of new ecosystems are disrupting our world and creating unprecedented opportunities and threats. Now we focus our efforts on proposing an effective methodology to not only understand these changes but also to develop an actionable plan for benefiting from the opportunities and for mitigating the risks found in this era.

The process of growing organizations requires management to make decisions about allocating resources that will be directed at scaling ideas and initiatives or terminating the initiatives before additional precious resources are exhausted. The CEO of a Fortune 500 company confidentially shared with us that the actual useable information reaching the senior leadership team is often missing a great deal of detail. This critical information that helps guide strategic decisions and resource allocation is simply not reaching the top decision makers. As described earlier, senior leaders are often not able to detect the early signals indicating approaching inflection points that typically surface on the edge of organizations. The problem is exacerbated by the fact that, once the leadership team is aware that there is significant detail missing from the reports reaching the top of the organization, it takes an extensive amount of time to gather the missing pieces, develop a complete analysis of the situation, and present a clear set of choices.

The Exponential Era: Strategies to Stay Ahead of the Curve in an Era of Chaotic Changes and Disruptive Forces, First Edition. David Espindola and Michael W. Wright.
© 2021 by The Institute of Electrical and Electronics Engineers, Inc.
Published 2021 by John Wiley & Sons, Inc.

The ability to effectively deal with rapid changes in a business environment is a function of the organization's decision cycle time and the comfort level with the information available at the time of the decision. In traditional strategic planning settings, these cycle times can span several years. However, these traditional planning cycles are no longer effective in an era of unexpected changes, technology convergences, and industry transitions that occur at exponential rates. The term "strategic planning" may even evoke images of stacks of paper collecting dust on the shelf of an executive's office. Regardless of how executives may feel about the term strategic planning, the fact remains that strategy and planning are executive functions that are even more essential in the Exponential Era. The problem is not with the label, but with the process itself.

"Plans are useless, but planning is indispensable."

DWIGHT D. EISENHOWER

Convergence Research and Strategic Planning

If traditional strategic planning cycles are no longer effective in an era of business changes and technology convergences that occur at exponential rates, how are executives to perform the essential strategy and planning functions that are still fundamental to their roles?

In an era where the pace of change is accelerating and new technologies are unleashed without warning, influencing the behaviors of people charged with executing a cohesive and effective strategy can seem like a daunting responsibility. What is needed is an equally fast, responsive, and information-rich process with early signals, as free of noise as possible. We need a methodology designed to deliver these signals in support of a decision-making process that can be repeated at speed, resulting in rapid scaling.

In order to develop a new approach to strategic planning, we have explored several methods and approaches that have effectively addressed the type of shortfalls we have identified in traditional strategic planning, such as long cycle-times and lack of flexibility. What we have found is a rich set of multidisciplinary methods and approaches that share common principles, and when uniquely combined offer the answers we were seeking.

One of our findings is that convergence is not only a driving force in the Exponential Era but also a positive aid in solving complex problems typical of this era, such as sociological changes in the Arctic due to climate change. Through convergence, the National Science Foundation (NSF) is finding solutions for the people living in the Arctic who are dealing with critical and urgent issues that require innovation from multiple sectors. They call it "Convergence Research," a way of bringing people together from various disciplines and backgrounds in what the NFS describes as "a deeper, more intentional approach to accelerating discovery." The tenets of Convergence Research include:

- Multidisciplinarity – Draws on knowledge from different disciplines while staying within the boundaries of those disciplines.
- Interdisciplinarity – Analyzes, synthesizes, and harmonizes links between disciplines into a coordinated and coherent whole.
- Transdisciplinary – Integrates the natural, social, and health sciences in a humanities context, transcending the fields' traditional boundaries.

According to Niki Wilson, author of the article *On the Road to Convergence Research*, "there is general agreement that convergence is not simply a set of experts coming together. Over time, team members absorb knowledge from one another and think with a different mindset, rather than just coming from the perspective of their own training and conditioning."[85]

By applying the same thinking behind Convergence Research, we developed an effective approach to strategic planning for the Exponential Era. We brought together a set of principles and methods from disciplines as diverse as software development, manufacturing, design, and the military. These principles and methods converge into an effective methodology for dealing with the challenges described throughout this book.

In developing Strategic Planning for the Exponential Era (SPX) we practiced exactly what we preach. We used lateral thinking to explore how other disciplines have dealt with similar problems. We went beyond our specialized domains to seek answers. We connected the dots in unique ways so that the whole becomes greater than the sum of the parts. We blended all these ingredients with our own knowledge and experience to create a new unique approach to strategic planning that accounts for fast, chaotic changes and disruptive forces characteristic of the Exponential Era.

The principles and methods that converge into our unique SPX methodology come from primarily four distinct but complementary disciplines: Agile, Lean, Design Thinking, and the OODA Loop. We will briefly describe each of them here in order to build the foundation for your understanding of our new approach to strategic planning.

The Manifesto that Changed How Software Is Built

Software development is a fertile ground for innovative approaches to managing the collaboration of several people who are trying to create a very complex, high-quality product. Software is prone to errors, or bugs in software parlance. The difficulty in translating business requirements into algorithms leads to constant changes. These changes are amplified by the fact that software is an abstract concept, and until you actually "see it and touch it," you don't really know if what you are getting is what you asked for. Worse yet, in many cases, you may not even be sure if you know what you want the software to do.

These difficulties cause many challenges in the management of software projects. Traditional software development follows a "waterfall" approach, which, in general terms, requires that the project proceeds phase by phase with each phase starting after its predecessor is complete. For example, no coding is done until all requirements have been fully gathered, understood, and documented. On the surface, this makes perfect sense. Software engineers are expensive resources, and the waterfall approach assures the project manager that no code will be written until the requirements are fully documented. This prevents the constant change problem described earlier.

This seems sensible at first glance. But the reality is that, as anyone who has ever been involved in a software project can attest to, what the software engineers understand to be the requirements, even if they follow very detailed requirements documents, are seldom what the business needs.

To address this challenge, several iterative and incremental development methods have evolved, culminating in the Manifesto for Agile Software Development, originally published in 2001 by several software developers that gathered together to come up with a better way to develop software.[86]

Agile software development is focused on people and their interactions. It places a high value on constant communication with the end customer or user. It accepts the fact that requirements cannot be easily translated from a document, and that constant, iterative change in collaboration with the end-user is key to developing working software.

The principles behind Agile software development can be summarized into continuous and early release of working, production-quality software based on daily cooperation between developers and end-customers. Changes are accepted as part of the process. In fact, the mindset of an Agile developer is that software is never truly finished, as it is constantly evolving. As long as you can maintain a constant pace of development, striving for technical excellence, good design, simplicity, and customer satisfaction, the software will incrementally evolve into a much better product than could have ever been envisioned and planned for in the initial phases of the project.

"We plan but recognize the limits of planning in a turbulent environment."

JIM HIGHSMITH (2001), "HISTORY: THE AGILE MANIFESTO"

Today, the principles of Agile have expanded into many disciplines that go much beyond software development. The idea of constantly iterating, listening to the customer, and delivering small increments that are highly adaptable can be used just about anywhere in business.

Iterative Experiments in a Feedback Loop

The next method that converges into SPX is Lean. Lean thinking was originally applied to how production systems and supply chains are run. Specifically, many of the ideas and principles found in Lean literature can be traced to a production method derived from the Toyota Production System (TPS) developed by Taiichi Ohno and Shigeo Shingo.[87] But just like Agile, Lean has a set of values and principles that can be applied to multiple disciplines.

The tenets of Lean include valuing individual knowledge and creativity, shrinking batch sizes, decreasing cycle-times, and removing waste. At a glance, you can start drawing parallels to Agile and its focus on people, continuous delivery of small batches of working code, reducing waste caused by excessive documentation and meetings, and fast, iterative cycle-times.

Applied to a broader context, Lean thinking is about the ability to effectively deal with change. Eric Reis captures the ideas behind Lean when applied to the startup world in his book *The Lean Startup*. Reis makes the point that instead of making complex plans that are based on a lot of assumptions, you can make constant adjustments with a steering wheel called the Build–Measure–Learn feedback loop.[88] The idea is that when you are operating in an environment where there are constant changes, as in a startup environment, you have to accelerate the learning process. You accomplish this by continuously iterating through the creation of prototypes, or Minimum Viable Products (MVPs), and measuring their acceptance in the market you are operating in (Product–Market–Fit).

The conversations you have with the end customers about your product will help you learn very quickly what is working, and what needs to be changed. What you are looking for is validation from customers themselves that the assumptions you have made about the product, customer needs, usability, pricing, and so forth, were correct. In most cases, you will find that you have missed the mark in at least one aspect of your initial plan, so you make adjustments and iterate again until you get it right. It is much easier to make these types of mistakes and adjustments when you are still dealing with a prototype or very early version of your product, and before you have made huge investments in product design, engineering, and marketing.

The Lean Startup methodology describes the startup's activities as a series of experiments that test their strategy and vision. It follows the scientific method in creating hypotheses that make predictions and then testing those predictions empirically to validate or discard the original idea, vision, and strategy of the startup.

Oftentimes, startup founders come up with an idea and subsequently put together a team in pursuit of that idea, only to discover after a series of experiments that the idea itself is not viable, but the team and its ability to execute on any idea are very valuable. Founders may at this point do a pivot, discarding the original idea but keeping the team to pursue other ideas that only became apparent during the iterative process of running experiments as part of the Build–Measure–Learn feedback loop.

This idea of running small experiments iteratively in a feedback loop is a foundational concept of SPX and is paramount to success in the Exponential Era where changes are happening at an increasingly faster pace.

Thinking Like a Designer

Next, we traverse to another discipline where, again, we find complementary ideas that converge into an elegant solution for dealing with strategy and planning in a fast-changing world. This time we move into the world of designers, and more specifically into Design Thinking.

Designers, like engineers, love to solve problems. But unlike engineering problems, design problems tend to be more open-ended, meaning, there is no right or wrong answer – there is only the design. Engineers can define a solution to a problem, test it, and solve it. The solution can then be repeated millions of times. Designers, on the other hand, deal with problems that are unique and that have no precedent.

There is no fixed or predetermined outcome to a good design. Designers have to deal with softer, abstract concepts like aesthetics, for which there is no right or wrong solution, only human emotions. This makes the job of a designer quite challenging. Designers have to deal with what some authors, like David Epstein, Bill Burnett, and Dave Evans, refer to as "wicked problems" – problems that are difficult to solve because they are incomplete, not well defined, contradictory, and with constantly changing requirements. These are exactly the types of problems executives have to deal with in the Exponential Era.

So how do you go about solving these types of wicked design problems? One method that has gained increased interest in the business press is Design Thinking. In fact, Design Thinking has gained so much traction in the business world that Stanford University launched a design school called d.school in 2005 to teach Design Thinking as a general approach to technical and social innovation.

The concepts behind Design Thinking can be applied to a diverse set of fields, from classic product design to the design of services, and even to life design. Bill Burnett and Dave Evans created a hugely successful course at Stanford University called *Designing Your Life*, which eventually turned into a book by the same title.[89]

In the book, the authors make the point that designers don't think their way forward. Designers build their way forward. They explain that designers don't just dream up great designs in their heads. They actually build things (prototypes) and try stuff.

IDEO is a design consultancy firm from Palo Alto, California, with strong ties to Stanford University. They were one of the first design firms to incorporate the concepts behind Design Thinking into their design processes. On their website, they describe Design Thinking as a human-centered core built upon three essential pillars:

- Empathy – Understanding the needs of those you design for.
- Ideation – Generating lots of ideas through brainstorming and other techniques.
- Experimentation – Testing those ideas with prototyping.[90]

Once again, we see the similarities, intersections, and complementary ideas coming from distinct disciplines that are trying to solve similar types of problems: how to deal with wicked problems that are hard to define, unpredictable, and constantly changing – the types of problems executives have to deal with as they develop strategic plans in the Exponential Era. Similar to the thinking behind Convergence Research, we believe that bringing people, ideas, and concepts together from these various disciplines and backgrounds will result in the best approach to solving these problems.

Getting Inside the Opponent's Head with OODA

Our journey through multiple disciplines in search of solutions to develop effective strategic plans in a fast-changing environment has taken us from the world of software developers to manufacturers, startups, and then to designers. Now we make one more visit to another arena, that of the military.

Here we introduce you to the concept of the OODA Loop, the cycle of Observe–Orient–Decide–Act that was developed by the military strategist and United States Air Force Colonel John Boyd. The concept was originally used in combat operations during military campaigns in order to gain an advantage over an opponent by quickly cycling through the process of observing and reacting to unfolding events, and "getting inside" the opponent's decision cycle to gain the advantage.

Just like we have seen with Agile, Lean, and Design Thinking, the concepts, and principles behind the OODA Loop have expanded beyond their initial origin in the military into other areas such as litigation, business, and law enforcement. The method emphasizes that the OODA Loop is a set of interacting loops that are to be used continuously – the key is to iterate in a fast tempo to stay ahead of the opponents. The four steps in the loop are:

- Observation – This is the part of the loop in which new information is being gathered continuously. The observation allows the person or organization to not only receive outside information as circumstances unfold but to also receive feedback from the other parts of the loop as they interact with the environment.
- Orientation – Here we use a set of filters such as culture, values, and beliefs to synthesize the information being gathered and feed the next step of the process. These filters will determine how we observe, decide, and act, making the orientation step the most influential.
- Decision – Here is where we decide what to do and formulate the plan of action. In making this decision we must take into account the synthesized information fed from the previous step in combination with our capabilities and risk predispositions. The result is a sequential, prioritized set of actions.
- Act – The actions taken will result in interactions with the environment that will unfold in sometimes predicted, but also in unexpected ways. This unfolding interaction with the environment is then fed right back to the first step in a continuous cycle.[91]

The cycle of Observe–Orient–Decide–Act continues iteratively, moving through a feedback loop that is always receptive to new information as it becomes available.

A New Approach to Strategic Planning

We find it fascinating that convergence, on the one hand, is creating wicked problems for business executives trying to build effective strategic plans in the Exponential Era. On the other hand, it has allowed us to find solutions by searching across disparate disciplines that deal with similar problems. The convergence of Agile, Lean, Design Thinking, and the OODA Loop, blended with our knowledge and experience, and connected in a unique way, resulted in the SPX methodology that we are proud to present to you here (Figure 4.1).

SPX is a completely new approach to strategic planning. It starts with the premise that it is possible to foresee transformative disruptions and map risks,

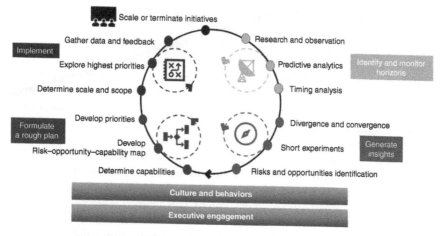

Figure 4.1   SPX: Strategic planning for the Exponential Era.

opportunities, and capabilities, allowing organizations to create and prioritize actionable plans designed to stay ahead of the exponential curve.

SPX, like the OODA loop, is a set of interacting loops that are used continuously. There are four distinct loops within the SPX continuous cycle, which we refer to as the SPX Flywheel, culminating in a decision to scale or terminate initiatives:

1 – Identify and Monitor Horizons
2 – Generate Insights
3 – Formulate a Rough Plan
4 – Implement

Under these four continuous loops, there is a foundational layer of Culture and Behaviors that is supported by a layer of Executive Engagement.

In the next chapters, we will examine each of the four SPX loops and the two foundational layers in detail. But before we dive deep into the details of the methodology, let's compare the traditional strategic planning processes that many of us have (grudgingly) lived through and that we have (loathingly) grown accustomed to.

Traditional Strategic Planning versus SPX

Traditional strategic planning is typically done over several months or years and culminates with an extensive document that lays out in detail the strategic vision and objectives of the organization for a period that typically spans two to five years.

In larger organizations, it is common to have an individual whose title is Senior Vice President of Strategy (or similar) who, aided by his staff, is ultimately responsible for the process that leads to the development of the strategic plan. This person works closely with the CEO and the senior executive team to set the agenda for the strategic planning process. This process may involve multiple meetings, numerous presentations, and several versions of the strategic planning document until a final version is agreed upon.

Every company does strategic planning a little differently, but for the most part, the process is the same. Visioning statements and strategic objectives are discussed in a series of executive meetings, new information is filtering through multiple layers of management, gap analysis is performed by the supporting staff, and results are reported up the chain until the final plan is locked in. This sets the stage for everything the organization will focus on for the next several years and establishes the budget for the strategic initiatives that will be pursued.

Traditional strategic plans consume a lot of resources and are very difficult to turn around once put in place. They reinforce organizational structures and business models that have worked in the past and make assumptions that what worked in the past is a valid predictor for what will work in the future. In this setting, it is difficult for executives to create a vision that reflects an in-depth understanding of a rapidly changing environment and the real challenges faced by their organizations as they get caught in the chaos of disruptive markets and new competitive forces.

This traditional strategic planning process may have worked well for companies that operated in industries that enjoyed stability for long periods of time. It made sense to establish long-term visions and objectives when operating in an environment where, short of an unexpected event, change happened at a moderate pace. This process is steeped in tradition and has stayed the same since it entered the business world in the 1940s. Its cadence has become so ingrained in the corporate world that it is seldom questioned, even though changes are now happening at a much faster pace, and it is practically impossible to envision what the business will look like in five years given that entire industries can sometimes be transformed in a fraction of that time, as we have seen happen to the newspaper and taxi industries.

Today, when no single industry is safe from being disrupted at unprecedented speeds, the traditional strategic planning process has become nothing more than an obsolete tradition languishing a slow death. Many of the initiatives drawn up from traditional strategic planning visioning statements turn into projects that eventually get shuffled in the mix of other priorities, as the day-to-day realities of a fast-changing world preoccupy the same people that are supposed to execute these projects.

While governance today is transforming, the strategic planning practices inculcated in most organizations are still anchored in the past. These can be harshly characterized as elaborate strategic planning processes in hermetically sealed boardrooms, isolated from the people who are facing the realities the business is dealing with in the frontlines. Leadership often promotes their newly established edicts in hopes that the troops will engage in following the new company direction. Often, after going through the motions a few times, employees become desensitized and apathetic. After the "all-hands meeting" they get right back to work, with little inspiration to change day-to-day behaviors, because if they don't make their quarterly numbers, they risk not getting a bonus.

SPX was designed to be drastically different from traditional strategic planning processes. We are living in an era where we can no longer assume that what has worked in the past will work in the future. This is an era where little can be taken for granted. In order to benefit from the vast opportunities afforded by disruptive changes while managing the many risks characteristic of the current environment, we need a new model for strategic planning that is much more responsive and agile than the models we have used in the past.

What differentiates SPX from past planning frameworks is its emphasis on detecting inflection points before they happen and making data-driven decisions that allow organizations to position themselves to enter the exponential curve just in time, before the Point of No Return. To help you understand the stark contrasts between SPX and traditional strategic planning, we summarized their differences in Table 4.1

SPX is a discovery-based process that favors adapting to conditions as they change. This flexibility contrasts with traditional strategic planning, which veers toward fixed scope initiatives that are typically broad, involving large teams, and costing tens, or even hundreds of millions of dollars. Instead of following a linear, low-frequency, long-duration, waterfall-based trajectory, SPX is iterative in nature. It is a series of short-cycle processes that repeat and adjust themselves at high frequency as conditions change and new information is gathered.

SPX posits that business conditions change fast due to the high impact of technology. It recognizes that in these situations the inputs to the process emerge rapidly and that factors that were known in advance may no longer apply to the new conditions. Therefore, it is paramount that leaders engage and communicate with all levels of the organization, especially the people at the frontlines who are most likely to detect the early signals that conditions are changing.

In such an environment, there is no time to write stacks of stale documents, so the outputs need to be dynamic, actionable living documents. As seen in our discussion about the progressions of the exponential curve, getting the timing right is absolutely critical. Start too early during the hype phase,

Table 4.1   Traditional strategic planning versus SPX.

Characteristics	Traditional	SPX
Scope	Fixed	Flexible
Approach	Linear	Iterative
Frequency	Low	High
Duration	Long	Short cycles
Business changes	Slow	Fast
Technology impact	Low	High
Organizational involvement	Low	High
Communication	Top-down, sporadic	Interactive, frequent
Input	Standard and known in advance	Emergent with rapid changes
Output	Fixed shelfware	Living document, actionable
Timing risk	Not considered	Multiple lenses
Response to uncertainty	Analysis paralysis	Experimentation
Implementation	Large initiatives	Iterative, fast execution
Requirements definition	Gap analysis	Immediate feedback

and you may waste resources and bet on the wrong platforms. Start too late and you may end up past the Point of No Return. Therefore, it is critical to observe timing to get it just right – more on that in Chapter 5.

Finally, SPX is all about responding to uncertainty. Unlike traditional strategic planning processes that become frozen by the inability to deal with insufficient information resulting in data gathering and analysis ad infinitum, SPX is an action-oriented methodology that relies on frequent experimentation to obtain the necessary data directly from the source. It supports quick decision-making based on what is known while maintaining the flexibility to adjust those decisions as new information becomes available. The key to SPX's effectiveness is iterating continuously through fast execution while making adjustments based on a feedback loop.

In Chapter 5, we will take a close look at the first loop of the SPX Flywheel: Identify and Monitor Horizons. This is the part of the process where the early signals are detected through both soft skills and hard science. We combine a human-centric, boots-on-the-ground type approach with in-depth research, predictive analytics, and timing analysis to detect the right signals at the right time.

CHAPTER 5

Detecting Early Signals

W e hope it has become abundantly clear from our previous discussions that traditional planning processes are no longer effective in the Exponential Era. We also hope we have impressed upon you how critical it has become to completely break from past traditions and to establish a new, innovative framework for effectively performing the essential executive function of strategic planning.

This new framework proposed by SPX emphasizes that organizations must detect inflection points on the exponential curve before they happen. Timing is absolutely critical here. You don't want to enter the exponential curve too soon when signals are so weak that you are unable to validate technology capabilities and market adoption, but you also do not want to enter too late, after opportunities from exponential changes have become obvious to all market participants, leaders have been established, and you are just a late entrant trying to play catchup.

The ability to identify and monitor "horizons" – the changing trends and technology convergences that can be hugely advantageous if caught early enough, or disastrous if too late – and to then create an effective plan that takes into consideration capabilities, opportunities, and risks, is what differentiates successful organizations from the Flash Boiled Frogs, and is the essence of SPX. In order to identify these horizons, you must detect the early signals, and in order to detect these signals, you must be listening.

For several decades NASA has been sending robotic spacecraft into the solar system to send back to Earth critical information that helps us get a glimpse of the wonders of our surrounding planets. These spacecrafts

The Exponential Era: Strategies to Stay Ahead of the Curve in an Era of Chaotic Changes and Disruptive Forces, First Edition. David Espindola and Michael W. Wright.
© 2021 by The Institute of Electrical and Electronics Engineers, Inc.
Published 2021 by John Wiley & Sons, Inc.

transmit signals that travel millions, even billions, of miles at very low power, usually at about the same wattage as a refrigerator light bulb (20 W). As the signal travels to Earth, it weakens. The signals arriving on Earth can be as weak as a billionth of a billionth of a watt – that is 20 billion times less than the power required for a digital wristwatch. In order to detect these tiny faint signals, NASA had to set up a system of antennas at three locations around the world called the Deep Space Network (DSN). To hear the low-power space-craft signal, receiving antennas on Earth must be very large, with extremely sensitive receivers.[92]

Analogous to how NASA has set up a DSN to detect faint signals that come from outer space, organizations must develop their own systems and mechanisms – a set of powerful, sensitive "antennas" – in order to detect the faint signals representative of the early stages of an inflection point, before they become obvious and strong. Furthermore, organizations must be able to distinguish the early signals that matter from the general noise they are constantly being bombarded with. This is no small challenge.

If It Is Obvious to Everyone, It Has No Value

Intuitively, detecting early signals that matter makes sense. But we would be amiss if we did not acknowledge that this is much easier said than done. In our decades of experience working with dozens of companies of all sizes across many industries, we can't recall a single organization or executive that would deny the importance of staying on top of technology trends and developments, or that would discount the validity of establishing systems and mechanisms that facilitate the detection of early signals. However, when confronted with the challenge of adopting a formal methodology that reinforces continuous horizon identification and monitoring through the detection of early signals, we have lost count of how many times we have heard an executive say: "We are already doing this."

Our formal consulting training and experience have taught us to listen carefully to clients, to ask well-thought-out open questions, and to always dig a little deeper into their answers in order to really understand what they mean. When we ask clients how exactly they are identifying and monitoring horizons today, and how they are detecting early signals, the answers can be easily predicted: they attend executive conferences and symposiums, they subscribe to technology journals and analyst reports, they belong to industry associations that provide ongoing updates on the latest developments. They have internal staff whose job is to know what is going on in their industries and be up to speed on related technologies and to report their findings up the chain. In other words, most executives we have interacted with do a very good job of getting educated and tuned in to what is being reported by industry consortiums, the mainstream media, and analyst firms. While we would encourage executives to continue seeking information through those channels, this is not at all what we mean by identifying and monitoring horizons.

To illustrate the point, let's look at a concrete example. Do you recall when you first heard about 3D printing? Most people recall hearing about it in the last 5–10 years, or somewhere in the 2010s. The first commercially available 3D printer was first offered as a kit in 2009, but it wasn't until 2012 that many different mainstream media channels and industry consortiums picked up on the technology. By 2013, we were already seeing significant growth and consolidation, like the acquisition of MakerBot by Stratasys, a current leader in the 3D printing market.[93] This was the point in time when leaders in the market were being established and the acquisition of surviving players was being made. The Point of No Return was solidifying.

Notice that the timespan between the first commercially available 3D printer and the Point of No Return was just four years. Most importantly, mainstream media channels did not pick up on the technology until 2012, just one year before industry leadership was being established. If you were an executive impacted by 3D printing technology, who, like many of the executives we have talked to, believed you were already identifying and monitoring horizons that impact your business by following the normal channels, you would have completely missed what today is a $10 billion industry, estimated to grow to $34.8 billion by 2024.

The reality is that the early signals for 3D printing were already detectable even before the year 2000, more than a decade before it became obvious through the normal channels. As an executive, if you had established effective mechanisms to detect the early signals, this would have been quite obvious. One of the many "antennas" that SPX advocates is the use of predictive analytics technologies to mine datasets that, when explored properly, can provide a wealth of information that can be used as early signals. One such dataset is freely available in the public domain: patent databases.

Had executives affected by the meteoric growth and wide-ranging impact of 3D printing used just this one mechanism recommended by SPX, they would have seen, for example, that the number of 3D printing patent filings in the United States was growing from the 1990s through the 2010s until it peaked in the 2011–2015 period as illustrated in Figure 5.1.

The peak of 3D printing patent filings in the early 2010s would have been an indication that the market was about to reach maturity, as it was later confirmed. But having had this foresight before it actually happened would have been a tremendous competitive advantage that would have guided strategic decisions at the right time. Using predictive analytics and mining patent databases to detect early signals in technology changes, are

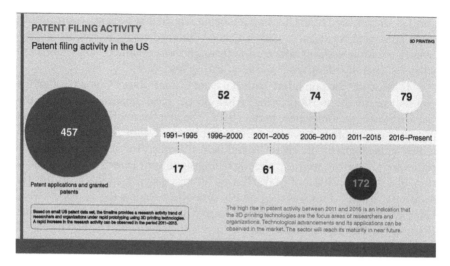

Figure 5.1   Patent filing activity in the United States for 3D printing per time period
Source: Intercepting Horizons.

some of the ways in which SPX helps you detect inflection points in an exponential curve – more on that later in the chapter.

When we talk about identifying and monitoring horizons, we are not talking about subscribing to analyst reports or mainstream media that report on what is happening today – these are current indicators. What we are talking about is seeking leading indicators. Being able to see 5, 10, or more years ahead so that you can detect the inklings of change and enter the exponential curve before opportunities become obvious to everyone else.

Many companies have a bias toward using current and lagging indicators to help them make strategic decisions. Financial metrics such as revenues, profits, and return on investment are lagging indicators. By the time a shifting horizon has had an impact on your financials, it is already too late. That horizon has already passed, and you failed to intercept it. Current indicators, such as the key performance indicators that you typically find on balance scorecards, are good at telling where you currently stand, but they are not of much help when it comes to telling you where things are going. In fact, current indicators can become blind spots as executives create incentive plans based on key performance indicators that focus on short-term results. They can also give executives the illusion that they are identifying and monitoring horizons, when in reality they are only seeing what everyone else already knows.

In order to detect inflection points before competitors, organizations must focus on leading indicators. But there is a problem. Some of the leading indicators advocated by some experts are based on suppositions, conjectures, and assumptions. In other words, they are not always factual. They are simply stories and narratives that require a lot of time and effort to process – we call them soft leading indicators. Soft leading indicators can be insightful, but most executives are uncomfortable and cautious about relying on nonfactual information as the basis for making strategic decisions. That is why SPX encourages using a combination of soft leading indicators with other techniques, including hard science.

Research and Observation

If you fail to detect the early signals while they are still weak or faint, you will not be able to create an effective strategic plan that enables you to enter the exponential curve before competitors have claimed leadership positions that are difficult, and many times impossible, to overcome. That is why the first loop in SPX – Identify and Monitor Horizons – is so critical.

Just ask Microsoft. According to Steve Blank, a well-known Silicon Valley entrepreneur, professor, and author, Microsoft missed the five most important technology trends of the beginning of the twenty-first century: They lost search to Google; smartphones to Apple; mobile OS to Google and Apple; media to Apple and Netflix; and cloud to Amazon.[94] Satya Nadella has done a phenomenal job in helping Microsoft regain lost ground in cloud, but they are still fighting a difficult battle for second place. As far as the other four technology trends and business opportunities are concerned, they have been lost forever.

Research and Observation is the first component of the first SPX loop, *Identify and Monitor Horizons*, as illustrated in Figure 5.2.

Research, by its nature, implies following a prescribed method to make discoveries, sometimes based on a scientific approach. Observation, on the other hand, can be done within the constructs of research, but can also be done through an informal process – we will discuss both cases.

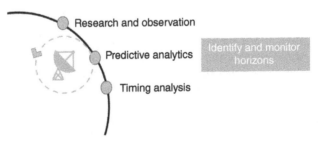

Figure 5.2 First SPX loop: identify and monitor horizons.

First, we approach observation from the standpoint that change first appears at the periphery and the first people to notice it are the ones in direct contact with it. Developing strategic plans from the safety of a corporate boardroom, being surrounded by internal staff, and without exposure to the periphery is not effective. Therefore, the first call to action in an organization's pursuit of leading indicators is to create mechanisms that open the flow of communication between the edges of the organization and the decision makers.

Removing Barriers to Information Flow

One such mechanism may seem simplistic but is surprisingly effective: the CEO spending a significant amount of time talking to customers and frontline employees. There are numerous examples of organizations that did not create such mechanisms and launched large initiatives without the prerequisite understanding of what was going on at the periphery, leading to disastrous results.

As described by Larry Bossidy and Ram Charan in their book *Execution – The Discipline of Getting Things Done,*[95] in 1997 Xerox hired Richard C. Thoman as COO in order to bring change. Thoman was a protégé of Louis Gerstner at IBM, where he had been CFO and was familiar with Gerstner's practice of spending his first months in office visiting customers and talking to frontline employees to get a deep understanding of what was really going on with the company. However, at Xerox, Thoman focused instead on launching numerous cost-cutting initiatives and was eventually promoted to CEO where he developed a new strategy to transform Xerox from a products and services company to a solutions provider. Thoman predicted that his strategy would grow earnings in the mid to high teens and that the company was "poised on the threshold of another period of great success."

However, Thoman's strategy was completely disconnected from reality. He launched two very difficult mission-critical initiatives that proved out to be more than the company could handle. The first was a consolidation of more than 90 administrative centers and the other a complete reorganization of the company's 30,000-person salesforce. These changes cast the company into a state of chaos, resulting in low employee morale, negative cash flows, and a plunging stock price that cost Thoman his job. What went wrong? According to Thoman's critics, he was too aloof to connect to the people who had to execute the changes.

In a similar account by Bossidy and Charan, Lucent Technologies named Richard McGinn CEO in 1996. An astute marketer, McGinn promised stockholders impressive growth in revenues and earnings. With the booming telecommunications equipment market at the time, Lucent seemed to be at the right place at the right time. But in reality, McGinn had difficulty getting things done inside the company. Even back then, technology was moving at extraordinary speed, but McGinn did not change the slow-moving and bureaucratic culture inbred from the Western Electric merger. Executives were flying blind without accurate information about customers, products, and channels – they had no way of making good strategic decisions and allocating resources accordingly.

The company missed the best emerging market opportunities and instead focused internally, wasting an enormous amount on a failed ERP implementation. The company's blind spots resulted in two fatal strategic mistakes: (i) It missed opportunities in the fast-growing network routers market by forgoing the acquisition of Juniper Networks and deciding instead to develop routers in-house, despite not having the capabilities to do so and (ii) It also missed opportunities in optical equipment due to its failure to understand the external environment. Senior management refused to listen to the early signals coming from internal employees who were close to the action and from customers who were seeing the shift to optical technologies. Instead, they focused on their biggest customers' immediate needs that were based on past successful products that were still profitable, but with poor future prospects, failing to foresee a shift in the market.

Adding to Lucent's mistakes were a number of misguided acquisitions setting out the company in too many directions and adding a myriad of unprofitable product lines and businesses that it could not integrate, creating redundancies, excessive costs, and a mass exodus of critical employees. Consequently, the company amassed a huge amount of debt that put it near bankruptcy. No one could imagine that a promising company like Lucent could fall so sharply. What went wrong here? According to published accounts, McGinn failed to probe his organization and get a realistic assessment of its market risks. During his last year in office, he was completely out of touch and in denial of the problems the company was facing. In October 2000, he was fired.

Clearly, both Xerox and Lucent failed to put in place the mechanisms that would have allowed them to capture leading indicators and avoid disastrous consequences. In particular, we see here two CEOs, who, despite the right pedigrees and many great qualities, were out of touch with both their internal organizations and the outside market. They were running their companies from their hermetically sealed corporate boardrooms, missing the evidence of change that emerges at the periphery. If both Xerox and Lucent experienced colossal falls due to their inability to detect early signals more than 20 years ago – a time when things were moving fast but nowhere near the pace we see currently – imagine what happens to companies in similar situations today.

By implementing mechanisms that systematize the flow of information from frontline interactions and from the outside to the corporate boardrooms, organizations can avoid the disastrous results and precipitous collapse experienced by Xerox and Lucent due to their inability to detect the early signals where they first emerge, at the edges.

Access Is the Name of the Game

Steve Blank in his classic startup book *The Four Steps to the Epiphany: Successful Strategies for Products That Win*[96] describes what he defines as a Customer Development Model. The book is focused on helping startup founders address a common problem that many entrepreneurs face: building products for which there is no demand. In order to address this typical blind spot, Blank has one powerful advice to entrepreneurs: Get out of the building! What he is getting to, is the fact that in order to understand customer problems, and what true demand may be out there to be solved by a startup's products, founders need to be in their customers' shoes, and truly experience a day in the life of the customer. It would be wise for executives in established companies to follow the same advice.

Startups learn quickly, sometimes through failure, how to access and sense early signals. Startup companies that are working on the wrong products, technologies, or platforms simply do not survive. But the ones that do survive can be a rich source of information for established companies. Many startups work on leading-edge technologies that may not be proven yet. By staying close to these startups through a variety of means, such as incubators, investment communities, or even in-house partnerships, mature companies can gain access and great exposure to early signals that they would otherwise miss.

Being where the action is, broadens an executive's perspective and deepens his/her understanding of the many external factors that may be influencing the company's products and markets. In order to remove the barriers to information flow, especially hierarchical barriers, executives can borrow from practices seen in companies that have successfully created methodical and organizationally safe ways to learn from the edges. For instance, at Citibank, leaders are asked to report on insights learned from actual customers regularly. Other companies randomly choose employees to have breakfast with their senior leaders.

Executives can also borrow from the Private Equity (PE) playbook. PE firms are tuned in to what is happening in the industries that they invest in, and they are always looking ahead, trying to find leading indicators. One of the ways they do this is by gaining access to industry and technology experts – the very people who are living day in and day out the challenges, competitive pressures, and opportunities that only people in the periphery would know. PE firms use service providers such as AlphaSights and Third Bridge to gain access to and connect with people that PE firms would otherwise not be able to tap into. In our view, it makes perfect sense for business executives outside of the PE industry to do the same.

In an era driven by digital technologies, ownership of physical assets no longer has the same weight it once had, since organizations are now able to access every asset class through licensing, rentals, and a variety of "as a service" subscriptions, from infrastructure to business applications and even pre-trained AI models. In the past, physical assets created barriers to entry, whereas today competitive advantage is achieved by the unique ways in which companies access advanced data capabilities through Application Programming Interfaces (APIs) and seamlessly integrate their core competencies into a cooperative ecosystem of partners. Ecosystems allow organizations to scale exponentially faster than they once could. Access has become much more strategic than ownership.[97]

Dealing with Uncertainty

Amy Webb points out that companies often miss opportunities and risks because they tend to have too narrow of a focus. She relates her experience as follows: "In my experience, companies too often focus on the familiar threats because they have systems in place to monitor and measure known risks. This adds very little value to long-term planning, and worse, it can lead to organizations having to make quick decisions under duress. It's rarer for companies to investigate unfamiliar disruptive forces in advance and to incorporate that research into the strategy."

When advising a telecommunications client dealing with uncertainties in their industry, Webb noticed that the types of exponential questions she would ask them was beyond the scope of their research. Even though they had a rigorous approach to planning, they tended to be narrow, with a focus on financial projections, competitive analysis, and technology assessments within their industry. She points out that what she observed is hardly unique: "When faced with deep uncertainty, teams often develop a habit of controlling for internal, known variables and fail to track external factors as potential disrupters. Tracking known variables fit into an existing business culture because it's an activity that can be measured quantitatively. This practice lures decision-makers into a false sense of security, and it, unfortunately results in a narrow framing of the future, making even the most successful organizations vulnerable to disruptive forces that appear to come out of nowhere. Failing to account for changes outside those known variables is how even the biggest and most respected companies get disrupted out of the market."

Webb suggests that organizations apply what she calls the "future forces theory," which explains how disruption usually stem from influential sources of macro changes, into their strategic thinking. She contends that all disruption can be attributed to 11 sources: Wealth distribution; Education; Infrastructure; Government; Geopolitics; Economy; Public Health; Demographics; Environment; Media and telecommunications; and Technology. She suggests that organizations look for areas of convergence, inflections, and contradictions, paying attention to emerging patterns and connecting the dots back to their industries, companies, and strategic plans.

Webb exemplifies her tenet by telling the story of two leadership teams that had access to the same "weak signals" that pointed to emerging changes in how people communicated. While one focused on trends within its industries and chose to make incremental improvements to its products, the other actively looked to broad external factors, connected the signals, and foresaw a world in which many technologies would converge into a single device. This resulted in the end of one of the world's most respected companies and the rise of an unlikely competitor. The former is Research in Motion (RIM), which at the time made the world's most beloved device – the Blackberry – which was so popular and addictive it was referred to in certain circles as the "crackberry." RIM at one point was one of the world's most valuable companies, with a market cap of $26 billion. Its narrow focus on incremental improvements such as changes in sizes and colors – what Clayton Christensen refers to as "sustaining innovations" – led to its demise. The latter is Apple, which transformed itself and an entire industry by introducing the concept of the smartphone, an innovative and disruptive idea that encompasses so much more than a simple communications device. Companies that insist on reinforcing existing practices and beliefs and that fail to broaden the scope of their research and innovation efforts are bound to face the same fate as RIM.[98] SPX was designed to address these challenges by helping companies deal with uncertainty and migrate away from outdated processes.

Conducting Research

The type of observation that we have discussed so far is informal in nature. It involves getting closer to the edges of the organization where the action is, or outside the four walls, in order to observe and learn from what is happening at the frontlines, in the world of the customer, and across industries. Next, we move to a more formal type of observation based on research.

Research, in the context of this book, is a data-driven discovery process that helps organizations uncover realities that may not be obvious. There is a wide range of research services available in the business world: from digital usability to quantitative market research; to focus groups; to industry analysis; to competitive intelligence and much more. Our goal here is not to describe each of these services in detail. You can easily learn more about them by doing your own searches or working with a specialized consulting firm. Our intent is to make you aware of diverse ways in which you can find leading indicators in order to gain competitive advantages in the Exponential Era.

Satya Nadella has done a remarkable job turning around Microsoft. As discussed earlier in the chapter, Microsoft missed the five most important technology trends at the beginning of the twenty-first century and was approaching the point of becoming irrelevant in the world of high technology. Today, thanks to Nadella, Microsoft has regained a prominent position as one of the world's most influential and valuable tech companies. How did Nadella accomplish such an implausible turnaround of this tech giant? In his own words from a 2015 interview: "We no longer talk about the lagging indicators of success, right, which is revenue, profit. What are the leading indicators of success? Customer love."[99]

If customer love is truly a leading indicator, as Nadella believes, then it is incumbent upon organizations to find out what customers are saying about them and their competitors. Are customers interested in their products? What are the demographic trends that may be impacting their business? These are some of the questions that quantitative market research can help answer.

One technique that is particularly popular in assessing "customer love" is the Net Promoter Score (NPS). NPS measures customers' overall perception of an organization's brand and is believed to be a leading indicator of growth. Customers are asked the likelihood of recommending the brand to friends and colleagues on a 0–10 scale. Then respondents are grouped into three categories: Promoters (score 9–10) are loyal enthusiasts who will keep buying and referring others; Passives (score 7–8) are satisfied but unenthusiastic customers who are vulnerable to competitive offerings; Detractors (score 0–6) are unhappy customers who can damage your brand and impede growth through negative word-of-mouth. Subtracting the percentage of Detractors from the percentage of Promoters determines your NPS. Knowing where you stand with your customers by conducting NPS surveys and tracking how it is trending can be a great way to determine whether your strategic plans are taking you in the right direction.

Another market research technique used frequently is focus groups. This type of research can also help you measure customer affinity to your products or brand, as well as many other research objectives. However, unlike the NPS, the type of information gathered in a focus group research is much more detailed, and the process is much more elaborate. Consequently, it can be more expensive but not necessarily more effective. Focus group research tends to yield varying results depending on many factors, but in particular how well it is conducted. The people selected to participate in the focus group need to be representative of your market, and the questions asked must be carefully crafted in order to retrieve valuable information.

Whether you are measuring "customer love" or other leading indicators, formal research, and its wide plethora of techniques, can be helpful in your horizon identification and monitoring process.

Predictive Analytics

The second element of the first SPX loop is Predictive Analytics. We use the term Predictive Analytics very broadly to designate the set of tools and technologies, such as data science, business intelligence, machine learning, and the broader Artificial Intelligence capabilities that organizations use to mine a variety of data sources in search of leading indicators. We chose this term not only because it is well-known in the business community, but also because the word "Predictive" captures the essence of what we are trying to do – make predictions about potential future outcomes – and "Analytics" because it is the means by which this is accomplished. However, it is also important to clarify that we are not suggesting that organizations can make accurate and precise predictions about the future – nobody can. What we are suggesting is that there are a set of tools and technologies, and a variety of accessible datasets, that allow organizations to recognize patterns and trends that can provide a range of potential outcomes and inform leadership.

As we consider how to use analytic capabilities to search for leading indicators, we broadly categorize two datasets that can be explored: internal and external. Internal datasets are your proprietary data that may reside in Customer Relationship Management (CRM), Enterprise Resource Planning (ERP), or other types of enterprise systems or data warehouses. This includes data about your customers and prospects that may be captured by your website, mobile apps, eCommerce, or any other lead capturing mechanisms in a sales funnel. External datasets refer to the public domain or commercially available data that you can access but that are not exclusively yours.

Let's start with internal datasets. CRM systems can be particularly powerful in helping you assess customer and prospect sentiments in order to measure trends in customer love as Nadella calls it. When we say CRM, we are referring to the broadest set of functions available in such systems, from capturing and responding to activities in social media; to customer assistance interactions online or through call-centers; to traditional sales forecasting and marketing. Social media has become the center of content marketing, and the information that can be analyzed through social media interactions can be helpful in determining how customers and prospects are reacting to your products and services as well as that of your competitors. Call centers and online customer service systems can also provide a very rich dataset that can be mined, sliced, and diced in many different ways. Do you want to learn more about what your customers think about your company's products and services? Just spend a day or two with your customer service reps. You will be amazed at how much you can learn. Do you want to know how those sentiments are trending? Data analytics running against CRM datasets will be a great place to start.

Artificial Intelligence engines running on top of your sales and marketing systems can help you predict future churn rates, sales conversions, campaign effectiveness, and much more. A detailed depiction of the specific ways in which internal datasets can be used to make a number of predictions is beyond the scope of this discussion. Suffice it to say that the tools and technologies available currently to help organizations capture leading indicators from the data they already own are many times more powerful than they were just a few years ago, and organizations are not using a fraction of this power to their advantage.

Most organizations are also not aware of external data sources that they can leverage to make predictions about future trends in technology. Once again, one of our favorites is patent databases. We showed you at the beginning of the chapter how executives could have predicted the emergence and eventual maturity of 3D printing by analyzing patent database filings. The same can be done for just about any technology you can think of. Stop for a moment and think about this. You could predict the emergence and maturity of just about any technology by analyzing freely available patent databases! And that is not all that can be learned from patent databases. As described in the Introduction, we have used Artificial Intelligence-based capabilities to analyze public domain patent data that led to the discovery of potential technology convergences that could completely disrupt and reshape an entire industry.

It is surprising how few companies use public domain patent data to gain insights into technologies and markets. Patent data can help innovators refine products, identify potential threats, perform competitor analysis, and develop Intellectual Property strategies to drive growth. Most importantly, it can be an extraordinary way for organizations to evaluate technological maturity, predict risks, understand market trends, and make investment decisions. We are so captivated by the usage of Artificial Intelligence and patent data to model predictions about future trends and convergences in technologies that

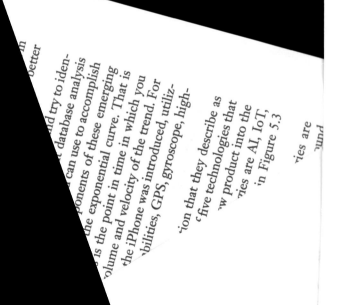

we have included it as one of the most valuable offerings in our portfolio of services. Now that the secret is out, we will probably see a lot more usage of this hidden gold mine.

To close the discussion about predictive analytics using external datasets, we would like to acknowledge the growth of commercially available solutions that organizations can leverage. The amount of data being collected every day is truly astounding. There are 2.5 quintillion bytes of data created each year and that rate is only going to increase as more and more devices enter the Internet of Things (IoT). In the last two years alone, we have generated 90% of all the data available in the world.[100]

With so much data being captured and made available for sale by both legitimate as well as nefarious sources, one word of caution is that going forward all companies must be careful about setting boundaries around data privacy and ethics. This is a topic of discussion in Chapter 10 of the book.

Timing Analysis

We now come to the third and last element of the first loop in SPX, Timing Analysis. Timing is absolutely essential for companies dealing with exponential changes. In fact, we would go as far as to say that timing is everything. We exemplified this in previous chapters where we considered companies that entered the exponential curve too early, and, more commonly, too late. If you enter the exponential curve too early, you run the risk of investing resources in platforms and opportunities that may not pan out. If you enter it too late, you may be forever playing catchup and not succeed.

If you recall from the Introduction, the key question in a Fortune 500 CEO's mind was "when." We all wish we had a crystal ball that would help us figure out the when of so many situations in our personal and business lives. Although we have not been able to find a crystal ball yet, we have found practical ways to deal with timing issues. We believe that by following a structured process that investigates early indicators of converging technologies and their timing, you can at least identify key markers on the exponential ~ guide your decisions.

The first step in this process is to ide~
the exponential curve with re~
trends that may ha~
are three key mark
Turning Point, and
the timing analysis
are still weak. What
nologies that will dr

they p~
all the rig~
create just the~
These are the~
when of their strategic~
nities or threats a business~
when they will occur. Timing a~
when significant risks might c~
dict when ready to act in order~
need to be examined how to detect~
opportunities.
Now that we have examined how to detect~
next loop in the SPX process.

capturing and responding to activities in social media; to customer assistance interactions online or through call-centers; to traditional sales forecasting and marketing. Social media has become the center of content marketing, and the information that can be analyzed through social media interactions can be helpful in determining how customers and prospects are reacting to your products and services as well as that of your competitors. Call centers and online customer service systems can also provide a very rich dataset that can be mined, sliced, and diced in many different ways. Do you want to learn more about what your customers think about your company's products and services? Just spend a day or two with your customer service reps. You will be amazed at how much you can learn. Do you want to know how those sentiments are trending? Data analytics running against CRM datasets will be a great place to start.

Artificial Intelligence engines running on top of your sales and marketing systems can help you predict future churn rates, sales conversions, campaign effectiveness, and much more. A detailed depiction of the specific ways in which internal datasets can be used to make a number of predictions is beyond the scope of this discussion. Suffice it to say that the tools and technologies available currently to help organizations capture leading indicators from the data they already own are many times more powerful than they were just a few years ago, and organizations are not using a fraction of this power to their advantage.

Most organizations are also not aware of external data sources that they can leverage to make predictions about future trends in technology. Once again, one of our favorites is patent databases. We showed you at the beginning of the chapter how executives could have predicted the emergence and eventual maturity of 3D printing by analyzing patent database filings. The same can be done for just about any technology you can think of. Stop for a moment and think about this. You could predict the emergence and maturity of just about any technology by analyzing freely available patent databases! And that is not all that can be learned from patent databases. As described in the Introduction, we have used Artificial Intelligence-based capabilities to analyze public domain patent data that led to the discovery of potential technology convergences that could completely disrupt and reshape an entire industry.

It is surprising how few companies use public domain patent data to gain insights into technologies and markets. Patent data can help innovators refine products, identify potential threats, perform competitor analysis, and develop Intellectual Property strategies to drive growth. Most importantly, it can be an extraordinary way for organizations to evaluate technological maturity, predict risks, understand market trends, and make investment decisions. We are so captivated by the usage of Artificial Intelligence and patent data to model predictions about future trends and convergences in technologies that we have included it as one of the most valuable offerings in our portfolio of services. Now that the secret is out, we will probably see a lot more usage of this hidden gold mine.

To close the discussion about predictive analytics using external datasets, we would like to acknowledge the growth of commercially available solutions that organizations can leverage. The amount of data being collected every day is truly astounding. There are 2.5 quintillion bytes of data created each year and that rate is only going to increase as more and more devices enter the Internet of Things (IoT). In the last two years alone, we have generated 90% of all the data available in the world.[100]

With so much data being captured and made available for sale by both legitimate as well as nefarious sources, one word of caution is that going forward all companies must be careful about setting boundaries around data privacy and ethics. This is a topic of discussion in Chapter 10 of the book.

Timing Analysis

We now come to the third and last element of the first loop in SPX, Timing Analysis. Timing is absolutely essential for companies dealing with exponential changes. In fact, we would go as far as to say that timing is everything. We exemplified this in previous chapters where we considered companies that entered the exponential curve too early, and, more commonly, too late. If you enter the exponential curve too early, you run the risk of investing resources in platforms and opportunities that may not pan out. If you enter it too late, you may be forever playing catchup and not succeed.

If you recall from the Introduction, the key question in a Fortune 500 CEO's mind was "when." We all wish we had a crystal ball that would help us figure out the when of so many situations in our personal and business lives. Although we have not been able to find a crystal ball yet, we have found practical ways to deal with timing issues. We believe that by following a structured process that investigates early indicators of converging technologies and their timing, you can at least identify key markers on the exponential curve that will guide your decisions.

The first step in this process is to identify where you currently stand on the exponential curve with respect to key horizons or emerging technology trends that may have an impact on your business. As we have pointed out, there are three key markers on the exponential curve: the Initial Inflection Point, the Turning Point, and the Point of No Return. Ideally, you should be performing the timing analysis before the Initial Inflection Point, while the early signals are still weak. What you want to be looking for is the convergence of key technologies that will drive the exponential curve, producing the crucial Initial

Inflection Point. If you find yourself beyond the Turning Point, you are late to the game and will have to do some catching up. This is not a good position to be in, but with significant effort and focus, it is still possible to take advantage of the opportunity and become competitive, as we have seen with Microsoft in the cloud business. If you are beyond the Point of No Return, you are better off finding other opportunities to go after.

Next, you want to review the emerging technology trends and try to identify at what point in time they might converge. The patent database analysis described earlier is one of the best mechanisms that you can use to accomplish this. Then plot on a graph when all the key components of these emerging trends will arrive at a single point in time in the exponential curve. That is your Initial Inflection Point, meaning this is the point in time in which you should expect to see an increase in the volume and velocity of the trend. For instance, mobile applications took off once the iPhone was introduced, utilizing the convergence of mobile broadband capabilities, GPS, gyroscope, high-definition display technologies, and more.

Let's suppose a business has recognized a situation that they describe as follows: Initial research has identified the confluence of five technologies that could result in a window of opportunity to introduce a new product into the market when these technologies converge. These technologies are AI, IoT, 3D Printing, Electric Vehicles, and Blockchain. The diagram in Figure 5.3 depicts the situation in a visual format.

We can see from the diagram that these different technologies are moving along an exponential curve in unique ways – some have been around

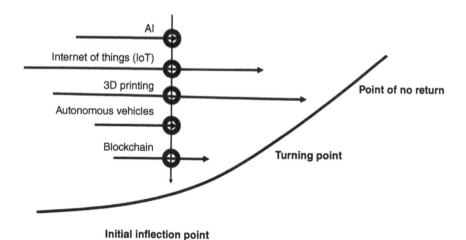

Figure 5.3 Timing analysis of converging technologies.

for a while, others have gone past the Initial Inflection Point, while others are arriving at the Initial Inflection Point synchronously. The observation of these technologies and their unique timing would guide the business to pursue answers to questions such as: How fast are these technologies growing? Have they peaked? What are their interdependencies? Is there a point in time when all the right conditions are in place for these technologies to converge and create just the right spark necessary for new opportunities and risks to arise?

These are the questions that will guide the business in determining the when of their strategic plans. What matters is not only what kind of opportunities or threats a business faces, or their sizes and consequences, but also when they will occur. Timing analysis allows organizations to potentially predict when significant risks might emerge, and how fast and how early they need to be ready to act in order to take advantage of time-sensitive opportunities.

Now that we have examined how to detect early signals, we move into the next loop in the SPX process.

CHAPTER 6

Learning from Experiments

U p to this point, we have focused our efforts on gaining access to and obtaining critical information to help us discern the changes that are happening that are impactful to our business and to help us make strategic decisions about timing. Now we turn to the next loop in the SPX process where our primary goal is to gain insights. This requires aggregating the information obtained thus far into a set of additional activities that will generate yet more information in order to give us a clear understanding of the opportunities and risks that the signals detected earlier represent.

Design Thinking will be of tremendous help in the first step of this process as we seek, like good designers, to first generate lots of ideas, and then selectively narrow down possible candidates to experiment with. As this process progresses, we will build hypotheses around technology changes, market shifts, social-economic trends, and other important factors that may impact our business and the risks and opportunities that they represent. These experiments will help validate or refute our hypotheses.

It is important to point out that, even though we have now entered the second loop in the SPX Flywheel, the first loop does not stop. The first loop will continuously listen to early signals while the second loop conducts experiments synchronously, as each loop delivers fresh input into the next, and provides feedback to its predecessor.

The second SPX loop, *Generate Insights*, contains three elements illustrated in Figure 6.1.

The Exponential Era: Strategies to Stay Ahead of the Curve in an Era of Chaotic Changes and Disruptive Forces, First Edition. David Espindola and Michael W. Wright.
© 2021 by The Institute of Electrical and Electronics Engineers, Inc.
Published 2021 by John Wiley & Sons, Inc.

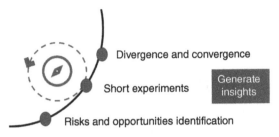

Figure 6.1   Second SPX loop: Generate Insights.

Divergence and Convergence

Divergence and convergence are concepts used in several domains, including math and social studies. In math, when we are moving toward a defined number, we are converging. On the other hand, when we are moving toward an undefined number or infinity, we are diverging. When we are talking about people, as in social studies or organizational behavior, convergence is the tendency of group members to become more alike over time. For instance, companies develop cultures which are the convergence of characteristics, values, beliefs, behaviors, and even the appearance of the employees. Divergence in this case is the tendency of group members to become less like other group members. For example, dressing differently in order to send a signal of your wealth, status, or nonconformance to societal norms. In the Exponential Era, we see examples of both. Technologies converge to create increasingly powerful changes to business and society. The problems created in this environment, however, are divergent. They are the types of problems commonly referred to as wicked problems, the sort that can't be easily defined, that are constantly changing, that have no prescribed answers, and therefore cannot be easily solved with solutions that have worked in the past.

In Design Thinking and in SPX, we diverge first, and then we converge. That is why this step in the process is called Divergence and Convergence, in this order, and not the other way around as you would normally find in most domains. The thought process here is that if you want to solve wicked problems, you first diverge, starting with an activity called ideation. Ideation is simply the process of generating lots of ideas in order to increase the chances that you will find an innovative solution to the problem you are trying to solve. Designers love generating a lot of "crazy" ideas because they understand that humans tend to be critical and leap to judgment very quickly. The ideation process allows us to defer judgment so that we don't inhibit the creation of great ideas that may be hidden in what initially appears to be just crazy.

Humans tend to fall in love with their first idea, but it is important to resist this tendency at this stage when the goal is to be creative and generate lots of innovative, wild ideas. A technique used in the ideation process is to jot down any ideas that come to mind, without censoring the words. Doing it fast will help contain judgment. According to David Kelly, Founder of the d.school at Stanford University, you often have to go through the wild ideas to get to the actionable good ideas. Going through the wild ideas is divergence. Getting to the actionable good ideas is convergence.

Starting with Divergence

In the last several years we have made significant advances in neuroscience, helping us get a still pale, but increasingly better understanding of the complexity of the human brain. The brain contains multiple complex structures that work together to help us function, each exercising an important role. For instance, the prefrontal cortex helps us make the types of judgment that we want to contain in the process of ideation, but that is essential for survival, as it prevents us from making foolish mistakes. The left anterior temporal lobe conducts semantic processing and helps us make sense of the world around us. Think of it as a filtering mechanism that becomes increasingly stronger as we age. Every new piece of information that we process is given meaning based on information that we have already retained. While this mechanism is valuable in helping us learn, it also narrows our view of the world based on our previous experiences, values, and beliefs. In other words, each of us has our own frames of reference from which we make sense of the world around us. But the same frames of reference may impart cognitive biases and may become an impediment to creativity and imagination. The problem becomes even more acute as we specialize in vocational fields, subjecting us to the old adage: "To a hammer, everything looks like a nail."

In his brilliant book *Range: Why Generalists Triumph in a Specialized World*, David Epstein extols the ability to take knowledge from one area and apply it to another by drawing on outside experiences and analogies.[101] We believe that this skill which Epstein refers to as "range" is one of the most valuable human skills in a world dominated by automation and Artificial Intelligence. Range and broad experience enhance our ability to perform human-centric pattern recognition – that intuitive, gut-based capability that machines lack. It is particularly effective in solving the ill-defined, fast-changing problems that contain few rigid rules that are not easily converted into an algorithm and that are commonly found in the Exponential Era.

Accessing knowledge that is not part of the normal channels and that resides outside of a company's four walls can be advantageous. Epstein calls it "The Outsider Advantage" and dedicates an entire chapter in *Range* to it. Epstein eloquently shows how people and entire organizations easily fall prey to the Einstellung effect, a psychology term for the tendency of problem solvers to employ only familiar methods even if better ones are available – also known as the "not invented here syndrome (NIH)." He also demonstrates that an effective way to solve wicked problems through innovative and creative solutions is to tap outsiders who use different approaches so that internal tribal knowledge does not bias and constrain the solution. He portrays how Alph Bingham, who was VP of Research and Development Strategy at Eli Lilly, used outside resources from diverse backgrounds to solve problems that perplexed Eli Lilly scientists. Bingham became so inspired by his experience that he decided to spin-off a separate company called InnoCentive. InnoCentive allows solution seekers in any field to post challenges that outsiders can solve in exchange for rewards. He calls the search for solutions in experiences outside of the problem's domain "outside-in" thinking.

Epstein quotes Karim Lakhani, co-director of the Laboratory for Innovation Science at Harvard, who researched InnoCentive solutions: "The further the problem was from the solver's expertise, the more likely they were to solve it." He adds further: "Big Innovation most often happens when an outsider who may be far away from the surface of the problem reframes the problem in a way that unlocks the solution." Some of the best ideas and paths to answers come from nontraditional sources like patent databases and experts from entirely different fields.

There are many solutions, like InnoCentive, that provide access to outside-in solvers in several specialized fields. For example, Kaggle focuses on solving problems related to machine learning. Interestingly, according to a top-ranked Kaggle solver, having knowledge of Artificial Intelligence techniques is not very helpful in solving the types of machine learning problems posted on Kaggle. Pedro Domingos, a computer science professor, machine learning researcher, and author of *The Master Algorithm* explained the reason to Epstein: "Knowledge is a double-edged sword. It allows you to do some things, but it also makes you blind to other things that you could do."

In another vivid anecdote, Epstein tells the story of Gunpei Yokoi, the influential figure at Nintendo that introduced the concept of "lateral thinking with withered technology" to transform the company from a seller of playing cards to a gaming powerhouse. Lateral thinking is about pulling together concepts from different domains into new contexts, so that old ideas find new uses. Withered technology is tech that has been around long enough to be easily understood and available inexpensively. Yokoi's inventive genius and ability to focus on the big picture and not be bogged down by engineering details, which was not his strength, resulted in the creation of the Game Boy, a smashing success that sold 118 million units, becoming far-and-away the bestselling console of the twentieth century.

The method Yokoi uses to find new uses for old technology is analogous to a technique that psychologists use in brain training exercises to improve creativity, called Alternative Uses Task. Trainers ask subjects to come up with new and original uses for an object, like a shoe, and count how many they can come up with. Through practice, subjects are able to improve not only the number of alternative uses they can come up with for random objects, but their entire creative process improves, as brain elasticity creates new pathways that increase the divergence of thinking and reduces what is known as functional fixedness – the tendency to consider only familiar uses for objects.

Yokoi understood functional fixedness and encouraged his employees to think more like producers and less like engineers, to avoid the trap of being lost in expert details and not seeing the forest for the trees. He also understood cultural pressures that prevented young engineers from voicing their opinions about novel ideas so that they would not look stupid in front of their boss. To avoid this problem, he would be the first to offer crazy ideas at meetings in order to set the tone and break the ice. As a lateral thinker himself, Yokoi understood the power of the ideation process and the importance of removing factors that might block the creative process that leads to truly innovative ideas. In other words, he encouraged divergence. But his geniality was also grounded on convergence, the ability to pull together existing ideas from multiple domains into a new idea. He believed that the best teams were composed of lateral thinkers, who could bring breadth, and vertical thinkers, who could provide expertise.

Moving into Convergence

Epstein's storytelling style leads to another fascinating account, this one close to home at Minnesota-based 3M. Andy Ouderkirk was an inventor at the company. For over 30 years he was named on 170 patents. In 2013, he was named Innovator of the Year by R&D Magazine. One of his greatest contributions was his multibillion-dollar invention of multilayer optical film that is used in a variety of products from cell phones to solar panels and fiber optics. A group of optics experts assured him his approach could not be done – to him this was a sign that he was on to something. Along the way, he became fascinated with the process of invention and teamed up with other researchers to study what profile of inventor made the greatest contribution.

Ouderkirk mined patent databases – as we have seen before, a great source for uncovering information about innovation – and wrote data analytics algorithms that discovered a type of inventor he called "polymaths." These inventors had depth in a core area but had worked across dozens of technology domains, allowing them to learn adjacent stuff while building breadth. This background was an ideal profile for the type of inventor who could converge the learnings from many domains into a new invention. They were also the most likely to succeed and win the Carlton Award, a kind of a "Nobel Prize" at 3M.

Ouderkirk believes that companies don't need too many specialists anymore because as communication technology improves and information becomes more broadly available, breakthroughs uncovered through narrow problem solving are quickly disseminated, becoming available to everybody. What is needed, according to him, is someone like Yokoi, who can converge these narrow technologies into applications.

Jayshree Seth, another corporate scientist at 3M who has more than 50 patents to her name, told Epstein that when she was a student, she had lost excitement for her area of research. She was considering changing her focus for her PhD in Chemical Engineering but was advised to stick to the area she knew but didn't like because she had already started down that path. This is an example of a concept called the "sunk cost fallacy," which we will discuss in more detail in Chapter 9. Seth decided to branch out regardless of the advice received and became what she describes as a "T-shaped person," one that has breadth compared to an "I-shaped person." What she became really good at is figuring out what questions to ask and converge the knowledge from the experts that know their specific stuff. She describes it as a form of mosaic building, like putting tiles together.

At one point in the past generalists lost favor to specialists. In the Exponential Era, the return of the experienced generalists will accelerate as company cultures change and recognize the value of range.

Returning to the SPX process, once you have accessed a diverse set of information across a wide variety of sources, including experts from many domains, you then need to turn to the aggregator. This could be an individual, like Seth, capable of integrating ideas from diverse experiences, or a team. The type of people you are looking to include in the team is what University of Utah professor Abbie Griffin calls "serial innovators." According to Griffin, serial innovators are system thinkers with a high tolerance for ambiguity. They are very good at repurposing knowledge that they gather from multiple domains and connecting them in unique ways. Their broad interests allow them to synthesize information that they aggregate by reading widely and talking to many experts in outside domains.[102] The aggregators play an enormously important role by converging all the information accessible and turning it into actionable next steps in the form of experiments.

As we continue to unpack the concepts and wide-ranging techniques that makeup SPX, we hope some divergence and convergence patterns are starting to become evident: Start by seeking diversity, gaining access to many sources of information, tapping outsiders, using lateral thinking, broadening your view beyond your specialized field or industry, and staying in touch with the edges. Then aggregate, connect the dots in new and wildly different ways, put the pieces together, creatively integrate multiple domains, and combine things in new ways to solve problems that are difficult to define, that change quickly, and that befuddle specialists. In short, find ways to beat wicked problems that will become ever more present in the Exponential Era.

Short Experiments

We now move into the second element of the second SPX loop, "Short Experiments." If you want to position yourself at any point in the SPX methodology, just refer back to Figure 4.1. Thus far we have collected a significant amount of information and synthesized it into a few ideas. These ideas are the ingredients used to formulate a hypothesis. Once the hypothesis is developed, it needs to be validated through experiments.

Notice that in SPX we qualify experiments by adding the adjective "short." This is done so that it becomes abundantly clear that we are not talking about large initiatives. Traditional strategic planning takes months to complete, leading to large initiatives that can span additional months or years. In contrast, these experiments should be very small and should start immediately. The goal is to prevent creating a tremendous amount of waste if the hypothesis is refuted.

We also want to make a distinction between the short experiments conducted here versus the implementations described in Chapter 8. All activities in SPX are experimental in nature but the experiments conducted here are shorter and may involve internal projects only, in order to spark learning or discover new possibilities. It is possible that you will run experiments in this stage that involve the customer, but unlike the activities in the implementation phase in the fourth SPX loop, at this point in the process, you typically do not have a complete offering yet.

Companies that have embraced a digital operating model create experimentation platforms that involve algorithms and data. For instance, Google runs more than 100,000 experiments each year to test a number of product features and potential improvements to existing products and services. Netflix has created an extensive experimentation platform that allows it to conduct A/B testing for all product changes.[103] The more digital a company becomes, the more the testing process can be automated, and the more the experiments can be scaled. Digital experiments can range from small cosmetic changes to strategic initiatives in target markets.

Because of the iterative nature of SPX, you will be conducting both short experiments and implementations simultaneously. Short experiments will result in prototypes and possibly Minimum Viable Products (MVP) that may or may not involve customers, whereas the implementations will always be live, with real customers, and complete offerings. Even though we don't discuss the latter until we get to the fourth SPX loop, be aware that all of SPX involves continuous iterative experimentations, simultaneously adding momentum to the SPX Flywheel. The key distinction is the short duration, prototype driven nature discussed here versus the live implementation of a complete offering covered in the fourth loop.

Before we start detailing the process of experimentation, let's discuss the hypothesis. You can think of the hypothesis as a set of your key assumptions. Assumptions are not facts – they are reasonable suppositions about the future. In the development of the strategic plan, it is important that the assumptions be clearly stated so that everyone knows what they are. Companies often get in trouble when assumptions are hidden, haven't been thought through, or are not well understood.

The hypothesis can be about a wide range of topics such as technology changes, new products or services, market shifts, customer adoption, and so on. For instance, you may be creating a hypothesis about the value a new product or service will deliver to your customers, or about the market's ability to discover, adopt, and use such products and services. Perhaps your hypothesis does not involve customers and markets yet. You may be at a stage in your innovation process where all you are trying to do is understand if the convergence of some technologies is promising, or if the unique use of concepts from other domains can indeed be combined into a new idea. Whatever your hypothesis may be, your goal is to validate it, refute it, or modify it through experimentation . . . quickly.

The most important advice to keep in mind in this process is having a bias to action. Remember, just like designers, you want to build your way forward. Designers try things, test them, create prototypes, accept failure as part of the innovation process, and keep iterating. It is possible that your original hypothesis will prove out to be wrong in the first few iterations, but that does not mean you have to throw it out altogether. Most likely you will have to continue refining your hypothesis until you find validation or the absence of viability. If you end up with the absence of viability, you will have to restart from scratch, completely dismissing your original hypothesis. Don't be discouraged when that happens. Think of it as a learning opportunity.

"I have not failed. I've just found 10,000 ways that won't work."

THOMAS EDISON

The next advice is to not anchor yourself to a bad solution. We have warned you about our human tendency to fall in love with the first idea. Don't fall in love with your solution. Instead, fall in love with the customer problem you are trying to solve.

If you find that you are stuck, one technique used effectively by designers is reframing. In *Designing Your Life*, Bill Burnett and Dave Evans tell the story of a 221-pound man named John whose dream was to take a mule trip from the top to the bottom of the Grand Canyon. The problem is that the mule can only handle 200 pounds. For years John tried to lose weight in order to fulfill his dream, but he could never quite get to the required weight. The authors point out that John could have reframed the problem from taking the Grand Canyon mule ride to seeing the Grand Canyon from top to bottom. He could then find a better solution than the one he was stuck with – losing the weight. He could have seen the Grand Canyon by helicopter, on foot, and so on.

The key is to not tie yourself to a solution that is not working and to reframe the problem in such a way that you can find other possible solutions. The challenge that most people deal with when they get stuck is that they tend to hold on to well-known approaches, the ones they are comfortable with, instead of breaking away completely from the familiar, and embracing the unfamiliar. Fears of the unknown can be minimized by trying to create small, safe prototypes that will not have significant consequences if they fail.

"Test fast, fail fast, adjust fast."

TOM PETERS

Your short experiment should take the form of a prototype or MVP. The MVP helps you learn quickly with a minimum amount of effort. The emphasis here is the learning process, not the actual product itself. What you are trying to do is simply validate your hypothesis. MVPs were not meant to be perfect, far from it. Polishing your product beyond what would be acceptable to an early adopter is a waste of resources and time. In fact, your MVP can be nothing more than a smoke test. In the *4-Hour Workweek*, Tim Ferriss suggests you test the demand for your ideas by placing online advertisements that take potential buyers to a landing page. If an order is placed, the potential buyer would see a new page stating that the product is not available. In reality, the product was not developed at all. The intent was to simply test the market demand hypothesis.[104]

Some MVPs are operated by people behind a curtain. As depicted in *The Lean Startup*, when Max Ventilla and Damon Horowitz founded Aardvark with a vision of creating a virtual personal assistant, they had humans replicating pieces of the backend for nine months. This allowed them to not only learn what type of questions people needed answers for but also bought them enough time to get the proper funding required to build the technology.[105]

Remember, one of the main objectives of the experimentation process is learning. One cautionary advice as you go through this process is to be careful that you don't cherry-pick the experiments and their respective results to fit your hypothesis. Having deep knowledge of a particular domain can be particularly dangerous here. As Pedro Domingos warned us, knowledge is a double-edged sword, and it can turn into a blind spot.

The leadership team at Netflix had long envisioned streaming as the ultimate technology for video consumption. But the real question was when to take action with respect to the anticipated inflection point. Rita McGrath explains in her book that Reed Hastings and his team created a small experiment to see if the time was right for his users to start consuming video via streaming – this part they got right. They launched the Watch Now service in January of 2007, offering about a thousand titles and charging $5.95 per month for a plan that provided six hours of streaming. Hastings enthusiastically declared: "We named our company Netflix in 1998 because we believed Internet-based movie rental represented the future, first as a means of improving service and selection, and then as a means of movie delivery."[106] But then Netflix took a number of missteps that led to a customer revolt with severe consequences. Those missteps could have been avoided through additional experimentation. It is possible that Netflix, despite good intentions, was too anxious to see the data fit their hypothesis and missed an opportunity to further experiment with business decisions that proved ill-timed. However, and this is what is ultimately critical in this story, Netflix learned their lessons and made several changes until they finally got it right. Today they are the undisputed leader in their domain.

The most curious scientists will always choose to look at new evidence, whether it agrees with their current beliefs or not. They understand that in a complex world, most cause-and-effect relationships are probabilistic, not deterministic. So, they dissect the results of experiments searching for lessons, looking not only at the evidence that validates their hypothesis, but also keeping a keen eye on evidence that refutes it. Learners rejoice in the lessons of a loss just as much as they cheer the validation of a win.

Risks and Opportunities Identification

We now move into the third element of the second SPX loop, "Risks and Opportunities Identification." So far you have identified and monitored horizons, diverged, converged, and run short experiments. What have you learned? What insights have you gained? What risks and opportunities have you identified? What have you learned about timing?

Any opportunity involves a certain amount of risk. Where one company perceives a risk, another sees an opportunity.

"Opportunity does not come gift-wrapped. You must take risks."

NINA BHATTI

In the Exponential Era, conditions are changing swiftly, creating turbulence that eludes common-sense. This age of chaotic changes, hyperconnectivity, and instant access to data (both legitimate and fake) many times generates confusion and fear. The primal brain kicks in and puts people in fight-or-flight mode.

As we write this, the coronavirus crisis is in full development. Surgical masks and hand sanitizers are flying off the shelves, events are being canceled, markets are plunging, and central banks are enacting emergency interest rate cuts. Is this a risk or an opportunity? Many companies have seen a significant drop in their stock price. The airlines and other travel-related industries are being hit particularly hard. Yet, Zoom Video Communications, the video conferencing company that provides an alternative to conducting business meetings without the need for travel has seen its stock price rise significantly in a few months, riding the coronavirus wave to an incredible $140 billion market capitalization.

In this environment, many investors are selling their equity positions in a flight to safety, fearing a coronavirus-triggered global recession, driving Treasury yields to record lows. Others are buying on the dip, based on a historical perspective that shows that epidemic-driven markets typically react in a "V" shaped manner. Risk or opportunity? We would suggest both, depending on your particular circumstances and perspective.

Embracing Chaos and Bad News

The strategic planning process advocated by SPX is not a tidy linear procedure devoid of mistakes and filled with neat little forms that help you document where you are, where you want to be, and the gap in between. SPX is not designed to help you deal with well-defined problems with deterministic outcomes. It is designed to address fast-changing, poorly defined challenges that are not necessarily well understood and for which there are no preceding formulae with right or wrong answers. It is messy, difficult, risky. It may seem chaotic at times. This is by design. McGrath points out that Andy Grove in his original work on strategic inflection points articulated that "before a pattern is clear you have to let a certain amount of chaos reign." It is through this chaotic process that risks and opportunities start to emerge.

McGrath also references experts like Ram Charan, Don Sull, and Nassim Taleb who agree that leaders should search for the presence of anomalies. They should be proactive in identifying risks and opportunities in the form of data or events that fall outside of the expected range. This requires communication practices in which total candor and brutal truth are the norms. Dissenting and conflicting input should be sought out, especially if it challenges the leaders' previously held assumptions – a belief and practice shared by some of the greatest American CEOs and management gurus of all time, including Andy Grove at Intel, Lou Gerstner at IBM, and Alan Mulally at Ford. In the process of seeking high-quality information, Taleb suggests a curious but astute approach: "Don't take advice from those who are not at risk."[107] Risk, when properly managed, can be a great ally. Sharing risk and rewards encourage stakeholders to get "skin in the game." When built upon a foundation of trust, it can be a very effective motivational factor.

Many traditional companies have formal risk management practices in their organizations. However, conventional risk management models have their origins from a time in which change happened at a moderate pace, so the focus was on the preservation of assets, rather than proactively seeking opportunities in what others might perceive as risk. Conventional risk management is about conducting audits, performing operational reviews, and managing compliance. In the Exponential Era, that is not sufficient.

Strategic risk management must encompass the idea that opportunity and risk are an inseparable pair. Risk is not something to be avoided at all costs but rather evaluated with respect to the accompanying opportunities. Risk can be mitigated through the process of experimentation, by making small bets, and by avoiding making large commitments that are not preceded by hypothesis validation and timing analysis.

In conclusion, the general goal of this SPX loop is to generate new insights that you would not have otherwise. As we have discussed so far and will continue to emphasize throughout the book, insight generation can be greatly enriched by extending your reach beyond your organization's four walls, your industry, and your domain of expertise. We encourage you to extend your teams by connecting to partners, academia, startup communities, subject matter experts, or just about anyone who can add value in the pursuit of applicable knowledge and insight. We call it the "network of networks."

Successful entrepreneurs are very good at this. They create vast networks that help them gather new information, recognize patterns, test hypotheses, and identify opportunities and risks. Through their network of networks, they are able to gain access to information that is not available through normal channels. They do this in real-time, and quickly leverage any newly gained information to make fast decisions, set up additional experiments, get more validation, and if needed, discard and rethink their hypothesis. This is the essence of insight generation and is a fundamental capability for companies that want to not only survive, but thrive in the Exponential Era.

Throughout the process of developing a strategic plan, organizations must be constantly evaluating risks and opportunities on an iterative basis as more data is gathered through the feedback loop, new hypothesis created, and new experiments run. The risks and opportunities identified in this stage then feed the next SPX loop, in which capabilities are considered in the development of a prioritization process that will eventually result in the formulation of an initial strategic plan. This drives the SPX flywheel momentum.

CHAPTER 7

Capabilities – The Essential Fuel to Ride the Exponential Curve

We have now arrived at the third loop of the SPX methodology called *Formulate a Rough Plan*. Why call it a rough plan? To make it abundantly clear that organizations don't need to spend months or years putting together a plan that will likely change as additional discoveries are made long before the end date of the plan.

We will continue to follow the same iterative process that we have described so far, taking the insights generated in the previous loop to help us make decisions about what to prioritize as we start to develop an initial version of a strategic plan. The previous loops continue to operate, identifying and monitoring horizons and generating additional insights. There are no waterfall steps here. All the loops are feeding each other continuously, so that no time is wasted as new information is learned, keeping the SPX Flywheel moving. Ultimately this loop will generate a prioritized list of undertakings the organization will pursue based on risks and opportunities identified earlier, taking into consideration capabilities that will be evaluated and developed in this loop.

The three elements of the third SPX loop, Formulate a Rough Plan, are illustrated in Figure 7.1.

We start by determining the organization's capabilities. As we discuss organizational capabilities, we want to make a clear distinction between the capabilities valued by traditional strategic planning processes and the capabilities called for by SPX. Traditional strategic planning and conventional risk management practices tend to place too much weight on maintaining stability, sustaining existing markets, and being compliant.

The Exponential Era: Strategies to Stay Ahead of the Curve in an Era of Chaotic Changes and Disruptive Forces, First Edition. David Espindola and Michael W. Wright.
© The Institute of Electrical and Electronics Engineers, Inc.
Published 2021 by John Wiley & Sons, Inc.

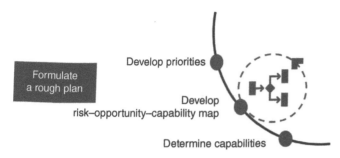

Formulate a rough plan

Develop priorities

Develop
risk–opportunity–capability map

Determine capabilities

Figure 7.1   Third SPX loop: formulate a rough plan.

This emphasis on stability worked in times of slow changes and incremental growth. But in the Exponential Era, too much emphasis on maintaining current operations and minimizing risk can result in a stagnated organization that clings to stranded assets producing revenue streams that are unsustainable.

The ideal organizations in the twenty-first century are ambidextrous. They are able to run current operations efficiently, while simultaneously being in sync with the disruptive changes around them. Most importantly, they are not afraid of foregoing efficiency or cannibalizing a profitable but dying business if that is what it takes to achieve sustainable, disruptive value.

The key questions to address at this stage in the SPX process are:

- Is the company's existence sustainable given the opportunities and risks identified previously?
- Does the organization have the capabilities required to not only survive but also to thrive given newly identified horizons? If not, how can the organization develop these capabilities?
- What is the risk tolerance of the organization?
- What risks is it willing to take, given the capabilities that it possesses or is willing to acquire or develop, to take advantage of horizon opportunities?

We will unpack these questions throughout the chapter, but fi- focus on understanding capabilities.

Determine Capabilities

Your goal in this step is to id
as well as the capabilities tha
we mean by capabilities? In

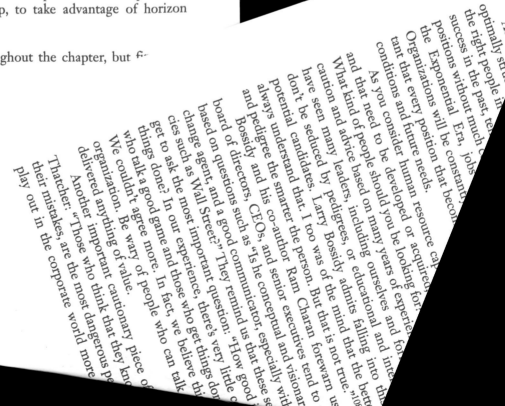

CHAPTER 7

Capabilities – The Essential Fuel to Ride the Exponential Curve

We have now arrived at the third loop of the SPX methodology called *Formulate a Rough Plan*. Why call it a rough plan? To make it abundantly clear that organizations don't need to spend months or years putting together a plan that will likely change as additional discoveries are made long before the end date of the plan.

We will continue to follow the same iterative process that we have described so far, taking the insights generated in the previous loop to help us make decisions about what to prioritize as we start to develop an initial version of a strategic plan. The previous loops continue to operate, identifying and monitoring horizons and generating additional insights. There are no waterfall steps here. All the loops are feeding each other continuously, so that no time is wasted as new information is learned, keeping the SPX Flywheel moving. Ultimately this loop will generate a prioritized list of undertakings the organization will pursue based on risks and opportunities identified earlier, taking into consideration capabilities that will be evaluated and developed in this loop.

The three elements of the third SPX loop, Formulate a Rough Plan, are illustrated in Figure 7.1.

We start by determining the organization's capabilities. As we discuss organizational capabilities, we want to make a clear distinction between the capabilities valued by traditional strategic planning processes and the capabilities called for by SPX. Traditional strategic planning and conventional risk management practices tend to place too much weight on maintaining stability, sustaining existing markets, and being compliant.

The Exponential Era: Strategies to Stay Ahead of the Curve in an Era of Chaotic Changes and Disruptive Forces, First Edition. David Espindola and Michael W. Wright.
© 2021 by The Institute of Electrical and Electronics Engineers, Inc.
Published 2021 by John Wiley & Sons, Inc.

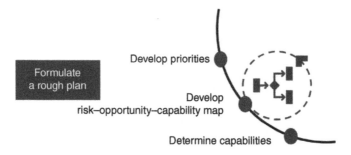

Figure 7.1   Third SPX loop: formulate a rough plan.

This emphasis on stability worked in times of slow changes and incremental growth. But in the Exponential Era, too much emphasis on maintaining current operations and minimizing risk can result in a stagnated organization that clings to stranded assets producing revenue streams that are unsustainable.

The ideal organizations in the twenty-first century are ambidextrous. They are able to run current operations efficiently, while simultaneously being in sync with the disruptive changes around them. Most importantly, they are not afraid of foregoing efficiency or cannibalizing a profitable but dying business if that is what it takes to achieve sustainable, disruptive value.

The key questions to address at this stage in the SPX process are:

- Is the company's existence sustainable given the opportunities and risks identified previously?
- Does the organization have the capabilities required to not only survive but also to thrive given newly identified horizons? If not, how can the organization develop these capabilities?
- What is the risk tolerance of the organization?
- What risks is it willing to take, given the capabilities that it possesses or is willing to acquire or develop, to take advantage of horizon opportunities?

We will unpack these questions throughout the chapter, but first, let's focus on understanding capabilities.

Determine Capabilities

Your goal in this step is to identify the capabilities that you currently possess, as well as the capabilities that you need to develop or acquire. But what do we mean by capabilities? In very simple terms we are referring to an organization's ability to do something well, sustainably, in order to create value for stakeholders. There are many ways to categorize capabilities, but we have chosen to group them in these five broad categories: people, process, technology, markets, and capital. Let's discuss each one.

People

When we say people capabilities, we are referring to its broadest meaning, both internal and external, including human resources, leadership, culture, knowledge, contractors, consultants, networks, communities, partnerships, crowdsourcing, and any other aspect of human-centric inputs and connections that the organization is able to effectively orchestrate.

Of all the capabilities an organization possesses, we would argue that people, especially internal human resources, and particularly the leadership team, are by far the most impactful in an organization's ability to thrive in the Exponential Era. It all starts with selecting the right people, and the higher the position in an organization's hierarchy, the more impactful that selection becomes. There is no higher priority for the Board of Directors than selecting the right CEO, and for the CEO and the entire leadership team to personally select the members of their teams.

We have pointed out examples of CEOs, like Richard Thoman at Xerox and Richard McGinn at Lucent, who despite many solid qualities, were out of touch with reality. We have also mentioned Reed Hastings, who made some mistakes, learned from those mistakes, and took the necessary actions to correct course and take Netflix to new heights. We also explained how Microsoft missed the most important technology opportunities of the early twenty-first century and had to replace the incumbent CEO with Satya Nadella who has been able to reorient the company toward prominence again. The person at the helm will set the tone for how the organization performs and will be the most influential factor in its ability to succeed.

But even boards themselves can be vulnerable, especially when it comes to being out of touch. We have all heard stories of boards that are not active enough in the affairs of the companies they oversee. There are numerous accusations of boards simply "rubber-stamping" the decisions of the leadership team and not probing enough to understand what is going on. Activist investors are increasingly using aggressive tactics to gain control of legacy corporations and to remove board members accused of passivity.

Active and competent boards will not hesitate to replace the CEO if the incumbent is not able to implement the necessary changes. One of the most important traits of successful leadership in the current state of fast technology changes and disruptive forces is decisiveness. Boards, CEOs, and the entire leadership team must be able to make difficult decisions swiftly and act on them. If a decision is made to replace the CEO, the next question is whether to leave in place the senior leaders inherited from a predecessor or build a fresh team. That assessment and the decision to reorganize must be done quickly. The reorg must be communicated promptly to let employees know that the leadership changes associated with the arrival of a new CEO are done and to build excitement around those who remain in the organization. The new organization must embrace the new directive and understand that they are now part of a new team embarking on a mission to implement changes and achieve new pinnacles.

At this juncture, an important question to ask is whether the company is optimally structured to achieve its goals. As the old analogy says: "Do we have the right people in the right seats on the bus?" Companies that have enjoyed success in the past, tend to create bureaucracies that automatically fill vacant positions without much consideration to the true need for those positions. In the Exponential Era, jobs will change significantly in a short period. Organizations will be constantly reshaping and reorganizing, so it is important that every position that becomes vacant be re-evaluated against current conditions and future needs.

As you consider human resource capabilities that you currently possess and that need to be developed or acquired, the natural question to ask is: What kind of people should you be looking for? Here we offer you a word of caution and advice based on many years of experience observing mistakes we have seen many leaders, including ourselves and former colleagues, make: don't be seduced by pedigrees, or educational and intellectual qualities of potential candidates. Larry Bossidy admits falling into this trap: "I didn't always understand that. I too was of the mind that the better the education and pedigree the smarter the person. But that is not true."[108]

Bossidy and his co-author Ram Charan forewarn us in *Execution* that board of directors, CEOs, and senior executives tend to evaluate candidates based on questions such as "Is he conceptual and visionary? Is she articulate, a change agent, and a good communicator, especially with external constituencies such as Wall Street?" They remind us that these senior leaders often forget to ask the most important question: "How good is this person at getting things done? In our experience, there's very little correlation between those who talk a good game and those who get things done come hell or high water." We couldn't agree more. In fact, we believe this applies to all levels of the organization. Be wary of people who can talk a good game but have never delivered anything of value.

Another important cautionary piece of wisdom comes from Margaret Thatcher: "Those who think that they know, but are mistaken, and act upon their mistakes, are the most dangerous people to have in charge." We see this play out in the corporate world more frequently than you could imagine.

Many people get into a position of authority through the power of their personality and their ability to influence people. These people tend to have big egos. They think they know more than they actually do, and their ability to persuade can be particularly dangerous if not kept in check. Smart company cultures counter this threat through an obsessive emphasis on data-driven decision making. Larry Page and Sergey Brin at Google are known for challenging their teams to back their claims with factual data in order to make a clear distinction between facts and opinions. Clearly, this obsession needs to be balanced with common sense. You don't want to create a culture so rigid that it inhibits people from expressing important inklings that may have not yet been verified by data but that deserve further investigation. "In God we Trust. Everyone else brings data" can sometimes be taken a bit too far as evidenced at NASA with the Columbia disaster. But the right dosage of skepticism and fact-checking can be effective in avoiding the problem that Margaret Thatcher alerted us to. Like most things in life, balance and perspective are key.

As we have discussed throughout the book, in a world where the solution to problems is no longer prescribed by known formulae, the people capabilities that you need to develop have shifted to a much more general set of knowledge, experience, and skills. Specialized knowledge accompanied by vertical, linear thinking no longer suffices in an environment that requires open lateral thinking and innovative solutions to wicked problems. You want people that can come up with new ideas, connect the dots in unique ways – aggregators who can identify the missing pieces. It is easy to be critical and point out problems that need to be corrected. It is a lot more difficult to raise ideas that have not been considered before, to offer a different perspective, to reframe problems, and to challenge the status quo. The people you want to attract and retain are collaborators who can demonstrate both depth and breadth. They are the polymaths, the lateral thinkers, the team players who can reach out to the experts wherever they are, inside or outside of your organization, to effectively combine ideas from multiple domains into unique solutions. You need aggregators in the Exponential Era.

Donald Rumsfeld's breadth of experience in both public service and industry as former Secretary of Defense and CEO of two Fortune 500 companies allowed him to collect an eclectic set of anecdotes, as well as various phrases, expressions, and snippets of advice that President Gerald Ford labeled "Rumsfeld's Rules." Rumsfeld eloquently shares them in his book appropriately titled *Rumsfeld's Rules*.

Teams must be diverse. However, as Donald Rumsfeld points out in his book, "the term diversity has accumulated a good deal of unfortunate baggage in recent decades. Too often, it can simply mean tokenism – diversity for diversity's sake." Diversity works, but it is not about meeting some arbitrary quota. It is about the diversity of thought that can only be achieved across several domains and experiences.

Rumsfeld relates his experience: "Some of the best ideas can come from the sparks and thoughts generated during lively discussions around a conference table or lunch conversation among people who have opposing views. A leader will often draw inspiration from such discussions – ideas he would not have discovered on his own."[109]

In addition to internal capabilities, organizations must extend their people capabilities to the outside world. Suppliers, contractors, and consultants are part of that world, and the same rigor applied in the assessment and development of internal resources should be applied to these outside resources as well. We have talked about the benefits of crowdsourcing through the use of services like InnoCentive to seek innovative solutions to problems that befuddle internal resources that are perhaps too close to the problem to be able to think outside the box. If you have not developed crowdsourcing capabilities yet, we would encourage you to start getting your feet wet. We would also encourage you to reach out to the community and build network capabilities. Get engaged with startup incubators to discover what new ideas startups might be working on that pertain to your industry or field. Connect with local universities to learn about leading-edge research that is being conducted and that might have the potential to be commercialized through partnerships with the private sector. There are numerous examples of successful companies that spun off from research conducted at leading universities that actively seek the commercialization of their patents.

People and their capabilities are an extensive topic, and as mentioned before, the most important for you to consider as you develop the initial version of the strategic plan. We will have additional discussions about people in Chapter 9 where we cover culture, leadership, and executive engagement in more detail.

Process

Process capabilities are fundamental to a company's operation. For our discussion, process capabilities are all the activities, procedures, policies, systems, automation, integrations, supply chains, manufacturing, warehousing, human resources development, finance, marketing, sales, or anything else that is part of the day-to-day running of a business.

In the 1990s, business process re-engineering (BPR) was all the rage. Companies spent gargantuan amounts implementing enterprise resource planning (ERP) systems that encompassed most back-office processes from Financials to Manufacturing, Supply Chain, Logistics, Transportation, HR, Product Lifecycle Management, and much more. Consulting firms, including the Big 4 Accounting firms, dedicated entire practices to implementing software from the top vendors such as SAP and Oracle. Originally, these projects were set up with a traditional waterfall approach, took years to implement, and cost clients tens of millions of dollars. The heydays of ERP implementations came before Agile concepts were widely adopted. One of the many shortfalls of these projects is that they started with a gap analysis. Armies of consultants would come in and document the current processes (the "As-Is"). Then they would interview stakeholders and ask them how they envisioned these processes working in the future (the "To-Be"). The resulting To-Be documents would turn into never-ending wish lists with process requirements so complex that no software in the world was capable of fulfilling them. However, the consulting industry had a magic solution in their back pockets that solved what appeared to be insurmountable challenges: "customizations."

More consultants with deep and expensive expertise in the intricate inner workings of ERP software would be more than happy to customize off-the-shelf software to meet their clients' sophisticated process needs under the justification that their clients' processes were "unique" and provided a competitive advantage.

The results were disastrous. Because of the deep and complex customizations, companies were unable to keep up with upgrades released by the software vendors that included software improvements, bug fixes, and additional security. These companies kept falling further and further behind with outdated software, and what was supposed to become a competitive advantage turned into process logjams that would prevent top-line growth and impact bottom-line results. Hundreds of projects were scrapped or redone. Millions of dollars were wasted.

The next wave of BPR came in the form of front-office process improvements with customer relationship management (CRM) software. By the time CRM became popular, the industry had learned from its ERP mistakes and agile approaches became more commonly deployed. CRM touched just about every front-office process from Opportunity Management to Forecasting, Quoting, Configurations, Incentive Compensation, Territory Management, Order Management, Field Service, Help Desk, Customer Support, Call Center, and much more. CRM implementations typically did not suffer from the same ills of their ERP counterparts, but process capability issues continued to plague many companies.

The problem now was the unanticipated creation of process silos that were not integrated and that in many cases were inconsistent across business units and geographies. Companies became enamored with "best of breed" CRM software from multiple vendors that promised to increase revenues by providing the latest and greatest capabilities that their competitors could not match. They targeted VPs of Sales who would do just about anything to make their salespeople more productive and reach their sales quotas. The result was a plethora of CRM systems that did not talk to one another, creating an unmanageable mess of processes and systems that, once again, promised competitive advantages, but delivered gridlock.

As companies started to increasingly make acquisitions as part of their growth strategy, the problems only got worse, as organizations were unable to deal with their process mess, let alone integrate another set of processes and systems that were nearly impossible to integrate. We called this plethora of disparate and disconnected processes and systems across departments, business units, and geographies "the quilt" – the graphical representation of the mapping of an organization's processes and systems looked exactly like a multicolor quilt, with each color representing a disconnected silo. This problem still plagues many companies today. Some companies have been able to consolidate and integrate many of their processes across the globe, but the problem is gigantic and widespread across many industries.

Given the historical perspective of the pursuit of BPR and process improvement delineated above, it is important to perform an honest assessment of your current process capabilities:

- Where are you at in your process capabilities journey?
- Are your processes integrated?
- Are they consistent across business units and geographies?
- Are you able to quickly integrate new acquisitions into your business?
- How is your customer experience?
- Do you have a 360-degree view of the customer?
- Is your call center able to address multichannel customer issues in one step, or are your customers being bounced around like a ping-pong ball?
- How much visibility do you have in your supply chain?
- Are your processes integrated with your suppliers?
- Are you able to provide your customers with accurate product delivery dates?

These are all fundamental process capability questions that every company must wrestle with.

Just to be clear about this, we are not suggesting that your strategic planning process turns into operational reviews and process improvement activities. What we are suggesting is that you must understand what your process capabilities are and what additional capabilities you must acquire in order to take advantage of the opportunities and mitigate the risks associated with the Exponential Era. Process capabilities are table stakes. If you don't have them, you will not qualify to play the game.

In addition to developing operating process capabilities, organizations must also embrace innovation processes. Innovation processes are embedded throughout SPX, so if you follow the methodology, you will adopt innovation processes that will position your organization to succeed in the Exponential Era.

Technology

In the Exponential Era, every business is a technology business. If you don't have technology capabilities, you are out of business. Technology capabilities include all the hardware, software, telecommunications, clouds, APIs, patents, licenses, documents, open-source, or anything else that makes up the total internal and external technology pools that you have access to and that you have the ability to use effectively.

People in Information Technology (IT) departments have a curious habit of referring to people outside of IT as "the business." That term never made much sense, but it is especially inadequate now that "IT is the business," as Martha Heller points out in her book *Be the Business: CIOs in the New Era of IT*.

As we have described in Chapter 2, technology convergences are creating new ecosystems that are changing business and society. Some technologies, like genome sequencing, are only pertinent to companies that operate in healthcare and adjacent domains. However, some technologies are ubiquitous across all businesses and industries, such as mobile devices, 5G networks, Artificial Intelligence (AI), cloud technologies, and the Internet of Things (IoT), to name a few. It has become paramount for businesses to assess their capabilities in these technologies and create a plan to acquire these capabilities if they are not already in place.

As we stated before, the impact of AI is particularly noteworthy as it becomes increasingly embedded in our everyday lives. Amazingly, and frighteningly, AI has the potential to understand what we want more than we consciously do. Gradually, we are relying on AI to make everyday decisions, not just the shortest route from point A to point B or how to avoid traffic and toll booths, but also what purchases to make and where. AI may deduce what we need to buy based on past purchases, current shortages, conversations that personal assistant devices are listening to, facial expressions, pupil tracking, and other indicators that AI may gather imperceptibly to us. Imagine the implications this new modus operandi will have on all consumer businesses and the broader implications to all business processes across industries.

The convergence of 3D Printing, Robotics, AI, and Autonomous Vehicles will provide businesses the ability to provide "instant fulfillment" deliveries of individualized products manufactured in route to their destination. This is a game-changer for all businesses that rely on logistics, transportation, and distribution to satisfy their customer needs.

We could list a myriad of technology changes and convergences that will completely transform businesses, but you can go back to Chapter 2 if you need a refresher. If you haven't already done so, the important point for you to realize is that your technology capabilities will dictate your ability to stay in business. The questions you need to answer are:

- What technology capabilities do you currently have?
- What capabilities will you need to acquire to address the future scope and scale of your business?
- What capabilities could you obtain very quickly and inexpensively?
- What capabilities should you develop internally or obtain externally?
- For the ones to be developed internally, what is the lead time to develop such capabilities, and what do you need to get started?

If you haven't started to develop a digital operating model, we strongly suggest that you begin soon. If your processes, software applications, and data are still embedded in individual and siloed organizational units, it is time to rearchitect. The best practice you can follow is the one used by Amazon. Create a software architecture that is modular and distributed. Use APIs to expose data and functionality. Do not allow direct access to data stores, direct links, or backdoors. Prepare all interfaces to be externalizable so that developers outside your four walls can use them. Preserve a common foundation, aggregate the data, and build an Artificial Intelligence factory that can process this data, experiment fast, automate decisions, and scale the business. Keep teams small and agile. You can find a much more detailed discussion about the digital operating model in *Competing in the Age of AI*, by Marco Iansiti and Karim R. Lakhani.[110]

We recommend that you perform a complete assessment of all your technology capabilities and create a plan to develop or obtain the capabilities that are missing based on your current view and understanding of the environment. But keep in mind that the environment is changing at a very fast pace, so again, minimize long-term commitments that are expensive and difficult to change, and focus on dipping your toe in the water in areas that look most promising regularly.

Market

A fundamental question to be considered is how well you understand your current market capabilities. Market capabilities are all the avenues, resources, and touchpoints with your customers. They include prospects, products, services, salesforce, sales channels, distributors, agreements, renewals, territory, or anything else that you can leverage to give you competitive advantages in the markets you participate in.

According to Bossidy and Charan, most businesses do not know their customers as well as they think. Buying decisions, especially for business-to-business customers, can be very complex, highly matrixed, and with long sales cycles. Having a deep understanding of the organization you are selling into, and the personalities and buying behaviors of the people making buying decisions is fundamental to achieving sales targets.

Organizations with mature market capabilities spend a significant amount of resources training and coaching their salesforce. They use sophisticated sales techniques and carefully craft their messaging and value proposition in a way that truly differentiates them from the competition. They may use persona targeted messaging to personalize their sales strategy and interaction with prospects according to their personalities. For instance, if your target decision maker is analytical, detail-oriented, and skeptical, you want to try to mirror that persona and provide them with fact-based detailed information about the benefits of your products backed by case studies, proof-points, and testimonials. On the other hand, if the decision maker is a big picture person who values relationships and relies on intuition to make purchasing decisions, the previous approach may come across as boring and off-putting and is not likely to be effective.

Every sales opportunity is different, and the best salesforces are highly trained in situational fluency. For example, they may adjust their sales approach depending on where the prospect is in their purchasing cycle. Have they already identified the need for a product similar to yours, or are they unaware of potential benefits and happy to maintain the status quo? Have they done their homework and understand your strengths and weaknesses versus the competition? Do they have an approved project and budget? Smart and effective salespeople know how to engage their prospects in conversations that help them determine three fundamental questions: Why buy? Why You? Why now?

As you assess your market capabilities you must consider growth opportunities in different market segments. A clear understanding and mapping of your market segments will help clarify what products and services you may need to develop, what channels to use, pricing strategies, and promotional activities appropriate for each market.

In their insightful book, *Play Bigger*, the authors argue that category is the new strategy. They posit that "A great message, a great product, a great innovation – these things are no longer enough on their own. Now it's critical to develop a great new market category in concert with building a great company and product." The idea is that companies that best frame the problem often define and take the category, becoming what they refer to as the "category king."[111]

The idea of positioning the company to be a category king is not new. Jack Welch was a great proponent of being number one or number two in every market a company operates in. What is proposed in *Play Bigger* that is different than traditional practices, is that you define a market segment as your own category. By defining and owning the problem, and establishing thought leadership around the problem, the market will start to perceive you as a leader in that particular space. Customers' affinity to solutions from leaders in a category, no matter how narrowly defined or niche-focused that category may be, will greatly benefit the category owners. Defining and owning your category is a great way to enhance your company's market capabilities.

In addition, consider your market capability strengths in terms of your ability to move products and deliver them to your customers:

- How strong is your distribution network?
- How quickly can you get products to your customers?
- In what geographies do you a have a strong presence, and what other geographies might you consider penetrating?
- Do you have the right channel partners to help you reach customers in areas where you are not present?

Also, take into consideration overall customer loyalty:

- What is your Net Promoter Score?
- What is your customer churn?
- Do you have the right focus on securing customer renewals?
- What trends are you seeing in product returns?
- Are you able to upsell and cross-sell to existing customers? Do they recognize and value your brand?

Understanding the competition is also fundamental, but companies sometimes miss the emergence of competitors who have more attractive value propositions. This is particularly true in the Exponential Era where fast-moving competitors with low-cost infrastructure and access to cheap technology can quickly appear out of nowhere. As you assess the competitive space, you are not looking for tons of data on what competitors have done in the past. What you need to know is what they are likely to do next:

- How will they serve their customer segments?
- How are they planning to increase market share?
- How will they respond to your new product offering?
- What acquisitions are they considering?
- Who could they partner with?
- Who else is out there in an adjacent space?

Another strategic consideration might be any internal capabilities that could be turned into a new market. Amazon Web Services is a classic example of repurposing an internal capability to create a new market opportunity. Amazon was able to not only compete directly with long time incumbents but also create a new ecosystem in the process.

These are all important questions to consider as you assess your market capabilities. Having an open and honest discussion about your market strengths and weaknesses will better prepare you to make prioritized choices as you consider risks and opportunities.

Capital

We finally arrive at the last category: capital. Capital refers to all the internal levers on your income statement and balance sheet that you can use to fund activities, as well as external access to capital through debt and equity markets.

Capital capabilities will determine the organization's ability to fund new projects and adapt to changing economic conditions while sustaining a balance of short-term and long-term goals. Companies need to be able to balance achieving short-term financial objectives while at the same time investing in extending the life of the business.

Bossidy recalls challenging a manager named Jerry who introduced a hockey stick plan where earnings dropped for three years while launching his long-term strategy: "Jerry, I can't have flat earnings for the company for three years, so who is going to make up the difference? If you want to engage in something that's got a substantial operating loss aspect to it, then it's incumbent upon you to explain how you're going to fill this so-called bathtub between now and when this project becomes profitable. If you can't overcome it, then the enthusiasm for investing in the project is diminished." Bossidy conveys how people can become remarkably creative if you make it clear that you are not going to give them an earnings holiday while the project is going, as he recalls Jerry's response: "I can take more profit out of this product line in the short term because I don't think its long-term potential is that good anyway. And I can sell off a small business and make a profit because I don't think it is the best business for us to be in. I can cut expenses by ten percent during this period as a way to generate more earnings. I can do four or five things that can overcome this loss from the new product."

As you assess your capital capabilities, consider each internal capital lever that you can use to fund activities. The Boston Consulting Group recommends four such levers: Revenue, Organization Simplicity, Capital Efficiency, and Cost Reduction. They suggest multiple tools for each one:

- Revenues: Revamp pricing model, improve salesforce effectiveness through training and customer segmentation, optimize marketing spend through data analytics;
- Organizational simplicity: Reduce the number of organizational layers to improve accountability, decision making, and operational agility;

- Capital efficiency: Reduce inventory, improve payables and receivables efficiency, sell assets, outsource functions, increase equipment efficiency, optimize project portfolio through prioritization and elimination of failed projects;
- Cost reduction: Decrease promotional spending, better manage suppliers, improve procurement, optimize logistics and network, streamline product portfolio, increase offshoring and reduce headcount, cut spending on travel, utilities, facilities, IT and services.[112]

Clearly, not all of these levers are applicable in all situations, but they give you alternatives that can be carefully considered in different scenarios. The key is to understand your internal capital capabilities and know when it might be appropriate to use each lever.

The other aspect of capital capability is access to external capital:

- Are you able to borrow capital through bank loans?
- What is your credit history?
- What assets can you use as collateral?
- Can you issue bonds in the broader debt markets?
- Can you attract equity investors?
- Are you a candidate for venture or private equity?
- How complex is your capitalization table?
- Can you access the public markets?

A thorough review of your internal and external market capabilities will enhance your ability to make critical strategic decisions.

People, process, technology, market, and capital capabilities combined allow organizations to deliver value to stakeholders. A holistic view of capabilities in light of potential risks and opportunities is a powerful tool in the selection of prioritized undertakings to pursue.

Develop Risk–Opportunity–Capability Map

Now that we have determined the organization's capabilities, we will move to the second element of the third SPX loop: "Develop Risk–Opportunity–Capability (ROC) map." An organization's ROC map will facilitate the development of a flexible priority list that undergoes constant review and iteration as horizons and timing shift. The objective is to determine how to take advantage of time-sensitive opportunities while minimizing risks, given the organization's current or future set of capabilities.

The first step in developing the ROC map is classifying your capabilities as high, medium, or low concerning the horizons you are evaluating. Horizons are points of departure denoting a significant change in the environment under study. Please note that we are using the term horizons, in this case, to generically represent anything you may be considering pursuing or undertaking that results in significant change, such as the launch of a product, a new R&D effort, a marketing initiative, etc. We will use the terms horizons, initiatives, and undertakings interchangeably in the development of the ROC map.

For example, let's say you have identified blockchain as a promising emerging technology that will have a significant impact on your business, and you are considering launching a new product that gives your customers total traceability to all the steps in the production and distribution of products they source from you:

- How would you classify your capabilities in reference to launching this new blockchain-based product?
- Do you have people with experience and knowledge in blockchain?
- How mature are your processes and systems as it relates to the ability to collect information throughout production and distribution?
- Do you have access to blockchain service APIs that you could use in your product or would you have to develop everything from scratch?
- Will your salesforce be able to convey a clear message and explain the benefits of traceability through blockchain technology?
- Do you have the capital to fund this project?

This capabilities classification of high, medium, or low is somewhat arbitrary, and you have to use your judgment as you consider how you assess your capabilities in the environment you are in. We are not looking for precise measures of capability, we are simply trying to determine orders of magnitude to help us make smart prioritization decisions.

The next step is to create a Cartesian coordinate grid with Risk on the X-axis and Opportunity on the Y-axis, creating four quadrants: Prioritize, Consider, Demote, and Discard, as depicted in Figure 7.2. The goal is to plot each horizon you are considering on this grid using a symbol that represents capability levels related to each of these horizons.

The "Prioritize" quadrant contains horizons that represent high opportunity and low risk. The "Consider" quadrant includes high-opportunity and high-risk horizons. The "Demote" quadrant comprises horizons that are low risk but that also represent low opportunities. And finally, the "Discard" quadrant holds high-risk, low-opportunity horizons. Once you have plotted each

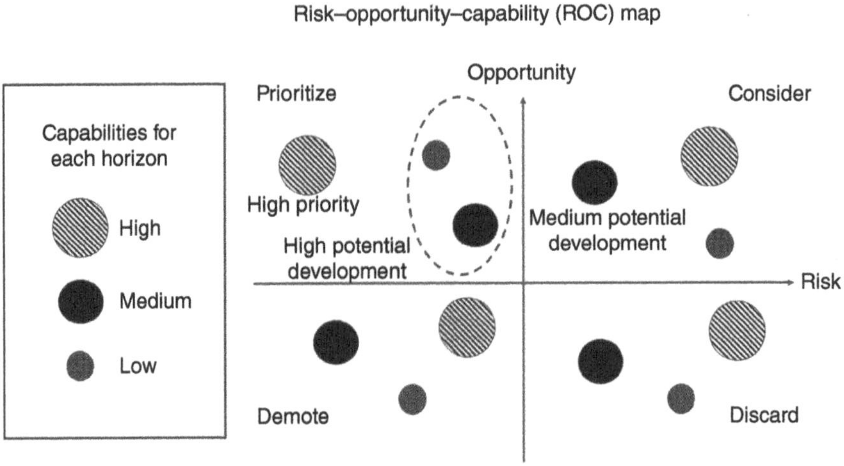

Risk–opportunity–capability (ROC) map

Figure 7.2   Risk–opportunity–capability (ROC) map.

of the horizons you are considering in the ROC map, you will have a great visual tool to help you create a priority list.

Develop Priority List

We finally arrive at the third and last element of the third SPX loop: "Develop Priority List." The goal here is to create a priority list of horizons to pursue and identify potential capability development areas in your organization.

We recommend organizing the priority list into high, medium, and low. To make this process easier, let's look at the ROC map and discard any horizons that will not make the priority list at all. Anything in the bottom half, in the Demote and Discard quadrants can be dropped. The horizons we want are in the top half, Prioritize and Consider quadrants.

We will start with the easiest choice by looking at the Prioritize quadrant. If you find horizons that have high opportunity, low risk, and high capability, they clearly belong in the high-priority list. But from there it gets a bit trickier. What if you have a horizon in the Prioritize quadrant for which the organization has medium or low capability? We call those high potential development horizons. These are horizons that you should pursue if, and only if, you can develop or acquire the capabilities required.

Next, let's look at the Consider quadrant. The decision here will depend on an organization's risk tolerance. As a rule of thumb, we recommend that you include all the horizons in the Consider quadrant for which you have high capability as medium priority. But you may also want to consider horizons in this quadrant for which you have medium capability. We call them medium potential development horizons. The same caveat applies here: you should only pursue these horizons if you can develop or acquire the capabilities required.

All other horizons on the upper quadrants should be considered low priority. To summarize:

- High priority: Prioritize quadrant with high capability
- High potential development: Prioritize quadrant with medium or low capability
- Medium priority: Consider quadrant with high capability
- Medium potential development: Consider quadrant with medium capability
- Low priority: All other horizons on the upper quadrants

We note here that the focus of the SPX methodology is identifying the best horizons for innovation pursuit and investment. Horizons that are dropped from consideration for this objective may still be valuable to an organization and could be monetized through other channels – e.g. licensing for high-capability opportunities that are in the "Demote" or "Discard" quadrants.

You have now formulated a rough plan of attack. There are still some pending questions, such as what capabilities you may want to develop or acquire. You also have to consider additional input from the previous loops and feedback from the next loop that will be generated iteratively and continuously. But now, armed with this prioritized classification of horizons to pursue, you are ready to move on to the last SPX loop where you will implement the rough plan that you have just developed, sustaining the momentum building in the SPX Flywheel.

CHAPTER 8

Feedback-Based Strategic Decisions

We have finally arrived at the fourth loop of the SPX methodology called *Implement* which will then take us to the last element of the SPX Flywheel. This last element, called *Scale or Terminate Initiatives*, stands by itself, and will also be covered in this chapter. These elements are illustrated in Figure 8.1.

The goal here is to implement and go live with the highest priority initiatives that we have come up with in the previous loop. To do that, we have to determine the scope and scale of each prioritized horizon that we would like to pursue so that we don't end up with more activities than the organization can handle. In other words, there is another level of selection and prioritization that needs to be done, but that can only be completed after we see the full picture of all contending horizons, their scale, and scope.

Before we delve into a discussion on scale and scope, let's pause for a moment to consider the fundamentals of strategy and the key elements of the strategic planning process. The word "strategy" is sometimes overused without much consideration of what it really means. Strategy at its core is a plan of action to achieve critical goals after careful consideration and prioritization of choices. Critical goals are ones with far-reaching consequences that will profoundly affect the direction of the organization. The development of an effective strategy requires careful planning, trade-offs, and difficult decisions.

Traditional strategic planning has gained a bad reputation due to its long drawn-out process that results in staid "shelfware." Organizations tend to put it off as they grapple with day-to-day demands, letting the urgent take precedence over the important. Strategic planning requires discipline, but when

The Exponential Era: Strategies to Stay Ahead of the Curve in an Era of Chaotic Changes and Disruptive Forces, First Edition. David Espindola and Michael W. Wright.
© 2021 by The Institute of Electrical and Electronics Engineers, Inc.
Published 2021 by John Wiley & Sons, Inc.

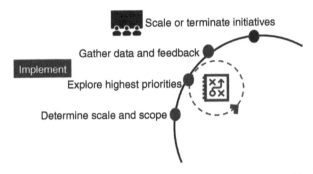

Figure 8.1   Fourth SPX loop: implement; scale, or terminate initiatives.

properly done, it becomes ingrained in the mindset and culture of the business. It is no longer thought of as an isolated event, as it becomes an integral part of the daily essence of doing business. Companies that do not embrace strategic planning as an indispensable discipline to be practiced continuously are driven by the inertia of outdated plans and operate in constant reactive mode. Their strategic reference point is the rearview mirror.

Strategic planning is not a set of meetings, a document, or an edict that comes from ivory towers. Just saying you have a strategy does not make it so. The strategy naturally unfolds as you work through the iterative steps prescribed by the SPX process. The result is a dynamic roadmap that orients the organization so that it knows what to do to reach its objectives. The roadmap is derived from a continuous process of prioritization and feedback that tells the organization whether it is headed in the right direction. Without a strategy, the organization is likely heading for failure.

Many clichés capture this concept, but like most clichés, there is often an element of truth in them. From the Cheshire cat in Alice in Wonderland: "If you don't know where you are going then any road will take you there." to Yogi Berra: "You've got to be very careful if you don't know where you are going, because you might not get there." But perhaps Sun Tzu said it best: "Strategy without tactics is the slowest route to victory. Tactics without strategy are the noise before defeat."

Setting the right priorities is a crucial step in the development of the strategy, and that is why we dedicate so much time and effort to this in SPX. Once the objectives and priorities are determined, the rest of the plan, if done correctly, follows logically. According to Donald Rumsfeld, this is also the most challenging part of strategic planning because it involves deciding among competing interests. In his book, Rumsfeld relates how British Prime Minister Tony Blair once stated: "The art of leadership is saying no, not yes." He also recounts Steve Jobs who said,

"People think focus means saying yes to the thing you've got to focus on. But that is not what it means at all. It means saying no to the hundred other good ideas that there are."[113] Deciding what you are not going to do is just as important as deciding what you will do, and that is the essence of developing a strategy.

"If you get the objectives right, a lieutenant can write the strategy."
                                                    GENERAL GEORGE C. MARSHALL

Donald Rumsfeld recalls many meetings in which he would ask senior officials what their priorities were. From their response, he would uncover the state of affairs. "If there was a hesitation on their part, it was telling. And if they rattled off a list of six or eight 'top priorities', that was telling as well."

Throughout this entire process, through careful consideration, you are trying to derive the three or four priorities that will drive your strategic plan. But to complete the prioritization process, we need to first determine the scale and scope of the undertakings being considered.

Determine Scale and Scope

We will now examine the first element of the fourth loop: "Determine Scale and Scope." At this stage in the process, deciding on the scale and scope of the prioritized horizons is critical. But what exactly do we mean by scale and scope? Think of scale as a vertical measure of depth, which is typically represented by quantities. How many customers will be impacted? How many transactions are expected to take place? What volumes will be produced? Scope on the other hand is a horizontal measure of breadth, which is typically represented by choices of inclusion. Which business units will implement it? Which markets will be served? Which departments will be involved? Scale and scope define the overall magnitude of the undertaking.

As we pointed out in Chapter 6, SPX is comprised of continuous iterative experiments. In the second SPX loop we conducted several experiments characterized by short duration with very small scale and scope that were geared toward the development of prototypes and Minimum Viable Products, whereas here we transition into the live implementation of complete offerings. You can think of it as transitioning from the Minimum Viable Product to what some authors refer to as the Minimum Lovable Product or Minimum Remarkable Product. The product can remain simple, but it must be complete, and most of all, customers must love it. The goal of these implementations is to try limited editions of each new undertaking we have considered thus far and gather feedback quickly. However, we need to keep in mind that some of these undertakings, even in limited editions, can take up a disproportionate amount of company resources.

Your challenge is to take each horizon on your priority list and determine its scale and scope:

- How many customer segments will be exposed to it?
- In what geographies will it be deployed?
- What parts of your organization will need to be involved and trained?
- What existing products and services will it touch?
- What integrations are required?
- How does it affect or change current sales and marketing initiatives?
- Does your overall messaging need to be refined?

These are just some of the numerous fundamental questions, along with many others, that need to be evaluated before taking action.

In alignment with our previous discussions on lateral thinking, as you deliberate the questions above in your scale and scope decision-making process, consider the following guidance from two adjacent domains: Lean and Computer Science.

Lean has taught us the benefits of using small batches. Taiichi Ohno, Shigeo Shingo, and others at Toyota revolutionized how cars are manufactured by introducing the concept of general-purpose machines that could be quickly reconfigured to produce a variety of parts in small batches. This was a completely new and innovative concept that contrasted with the practices of American manufacturers that used what appeared to be efficient mass production machines to manufacture large batches.

Although unintuitive at first, small batches are actually faster and more efficient than large batches because they don't require the extra time necessary to stack, sort, and move large quantities of parts. Problems can be discovered faster, and if changes are required, large quantities of inventory are not wasted. Large batches may seem more efficient from the point of view of the individual as it promotes specialized skills that in theory produce more through uninterrupted work. But in reality, when it comes to overall system performance, the flexibility afforded by small batches far outweigh the perceived efficiency of specialized large-batch production.

The lessons learned from manufacturing can be readily applied to other business concepts. In our case, the goal is to minimize the expenditure of time, money, and effort that could be wasted. Eric Ries, in this book *Lean Startup*, introduces the concept of the "large-batch death spiral." He explains that large batches tend to grow over time: "Because moving the batch forward often results in additional work, rework, delays, and interruptions, everyone has an incentive to do work in ever-larger batches, trying to minimize this overhead. This is called the large-batch death spiral because, unlike in manufacturing, there are no physical limits on the maximum size of a batch."[114]

As batches become larger, and releases take longer, there is a temptation to introduce just one more feature or fix just one more flaw. This reduces or constricts the flow of the process. Too much delay or constraint results in the flow stopping, leading to potential "death." Ries recalls his experience dealing with this problem: "I worked at a company that entered this death spiral. We had been working for months on a new version of a really cool product. The original version had been years in the making, and expectations for the next release were incredibly high. But the longer we worked, the more afraid we became of how customers would react when they finally saw the new version. As our plans became more ambitious, so too did the number of bugs, conflicts, and problems we had to deal with. Pretty soon we got into a situation in which we could not ship anything. Our launch date seemed to recede into the distance. The more work we got done, the more work we had to do."

Besides Lean, the other discipline we would like you to consider as you decide the scale and scope of potential undertakings is Computer Science. Computer Science provides powerful insights when it comes to the prioritization of large versus small tasks. There is a simple optimization algorithm called "Shortest Processing Time" that suggests you should always do the quickest tasks you can first. Christian and Griffiths explain in *Algorithms to Live By* that this is compatible with the recommendations made by David Allen in *Getting Things Done* to immediately perform any task that takes less than two minutes. They recognize that in business not all tasks are created equal, as some are more important than others, so they suggest capturing this difference of importance in a variable known as "weight." Here they offer a slight modification to the Shortest Processing Time algorithm: "Divide the weight of each task by how long it will take to finish, and then work in order from the highest resulting importance-per-unit-time to the lowest." If this seems complicated, here is an easy rule of thumb to remember: only prioritize a task that takes twice as long if it is twice as important.[115]

As you consider the scale and scope of the undertakings you will pursue, remember the Large-Batch Death Spiral and the modified Shortest Processing Time algorithm.

Also, consider the fact that what is possible today through Artificial Intelligence (AI) and digitization is considerably more scalable than what could be achieved in the past. These AI-driven processes easily connect with countless other digitized businesses, creating astonishing opportunities for learning and for producing ever more complex and accurate predictions. In essence, AI is transforming what it means to do business, enabling digital scale and scope, and removing deep-seated barriers that have constrained what companies were able to achieve for many decades. In an AI and automation-driven business world, scale and scope take up a whole new magnitude that was unthinkable just a few years ago.[116]

The choices you make about scale and scope will impact your prioritization process. You may have to reprioritize a few times as you consider other undertakings that might have been at a lower priority at first, but that due to their small scale and scope, should be given a higher priority. This entire prioritization process may become a bit untidy at times, requiring a lot of give-and-take, especially as new information is input into the process. But don't make it more complicated than it needs to be. Don't build a multitude of factors into your prioritization model.

Computer Science, once again, can be very insightful here: Statisticians and machine learning researchers often run into a problem called "overfitting." Research has shown that in prediction models it is not always better to take a greater number of factors into account. The extra factors not only offer diminishing returns – they can provide dramatically worse results.

At the end of this process, you should end up with a small reprioritized list that you are ready to implement. If you end up with more than three or four, you should reconsider. Remember that long lists don't get done. Instead, they stress the organization and demoralize teams when things don't get accomplished. Also, remember that if everything is a high priority then nothing is a high priority.

> "If you don't know what your top three priorities are, you don't have priorities."
>
> DONALD RUMSFELD

Before we move on to the actual implementation of your highest priority items, we want to leave you with a last word of caution about prioritization. Don't confuse enthusiasm with priority. A great and exciting idea can easily grab the team's attention – also known as chasing the shiny object. But just because the idea is exciting it does not mean it should be prioritized, or even executed. It is important to follow the disciplined process described here and only consider the factors that really matter in your prioritization decision: opportunities, risks, capabilities, scale, and scope.

Explore Highest Priorities

At this point, you have finally decided on the three or four priorities that you will deploy immediately. At last, you are ready to implement and go live. This takes us to the second element of the fourth SPX loop: "Explore Highest Priorities."

The term "explore" was carefully selected here. Yes, you are going live with real customers, and you may impact many people across your organization, but it is critical to remain flexible at each stage of the SPX methodology as you continuously gather feedback. Nothing is ever set in stone, and if the signals being gathered are not positive, or if the environment shifts and suddenly you realize you were moving in the wrong direction, you should not hesitate to shift gears, pull the plug, and start all over again. The ability to change course based on changing circumstances is a significant competitive advantage.

The benefit of strategic planning goes beyond creating a roadmap. We all know that roadmaps are rarely followed as designed. The most value comes from thinking through contingencies and preparing to adjust as unexpected circumstances arise.

"No plan survives first contact with the enemy."
HELMUTH VON MOLTKE THE ELDER

On the other hand, it is important that you give the strategy enough time to yield results. Seth Godin suggests that sometimes you might have a solution that people want, but you need to stick it out long enough so that people come to trust you. According to Godin, the longer you are in the game and the more time you spend with potential customers, the better your sense for your market becomes, leading to better decision making. You develop what he calls "practical empathy."[117]

In the early exploration of the highest priorities, organizations should not be worried about operational efficiency or scalability. The primary concern should be to discover if you are doing the right thing and if you are going in the right direction. The process may be full of inefficiencies, redundancies, and people behind the curtain. It is Ok to do things that don't scale right now. Your solution should be complete, but you don't need to have operational efficiency and scalability for your solution to be complete. You can worry about those things once you have confirmed that you have found market acceptance.

"There's nothing worse than the wrong things done right."
PETER DRUCKER

Do expect a reaction from your competitors. They will be watching what you do and will react accordingly. This may force you to revisit assumptions and reevaluate possible courses of action. Just when you thought you had a leg up on your competitors, they may introduce a superior competitive product that you could not have anticipated. However, we do not encourage too much competitive monitoring as it tends to distract from the key focus of disruptive innovation.

In addition, you should ask yourself the question: "If our customers could not find a solution from us, or from our competitors, where would they look for alternatives?" This simple question may help you discover possible new entrants and new technologies that might surprise you.

There may also be other unanticipated events that can catch you totally by surprise. Nassim Taleb refers to them as "Black Swans." In his bestseller book, *The Black Swan: The Impact of the Highly Improbable*, Taleb discusses these events, positive or negative, that are deemed improbable and yet cause massive consequences.[118] For example, as of this writing, we are living with the broad and unanticipated economic consequences of the coronavirus. Don't assume that just because you carefully thought through the strategy, detected early signals, experimented to test your hypothesis, and prioritized meticulously, you will achieve tremendous success. Always be prepared to deal with the unknown, and to adjust course as necessary.

As you implement the top priority initiatives, the initial contact with the "real world" will provide the signs that you need in order to decide whether to stay the course and scale or change directions. The most important aspect of this exercise is the learning the organization will go through. The implementation of the top priority items on your list will enhance your ability to detect early feedback signals for emerging trends before others experience them, but for this process to be effective you must be listening.

Gather Data and Feedback

Strategic planning does not end with implementation. A crucial aspect of the strategic planning process is to monitor progress through a set of defined metrics so that you can evaluate how well the plan is going and whether adjustments need to be made. This takes us to the third and last element of the fourth SPX loop: "Gather Data and Feedback."

The metrics that you define set expectations of what you are trying to achieve and are a powerful tool that must be used with care as it influences behavior. If you measure the wrong thing you will get the wrong results. Organizations are known for carelessly defining metrics that incent the wrong behavior. Call centers that measure time to resolution incent short and ineffective conversations with customers. Sales teams measured on call volume result in calls that don't lead to sales. The examples are numerous. So be careful with the metrics that you use to measure the effectiveness of your strategy.

"If you can't measure it, you can't manage it."

PETER DRUCKER

The best feedback comes directly from your customers and prospects, or the market in general. Their reaction, enthusiasm, and most importantly, their dollar commitment, best measured as total margin contribution by customer, is the best indicator that you are going in the right direction. However, there are a few caveats that you need to keep in mind as you seek feedback.

First, customers don't necessarily know what they want. Steve Jobs shrewdly warned us of this fact: "Some people say, 'Give the customers what they want.' But that's not my approach. Our job is to figure out what they're going to want before they do. I think Henry Ford once said, 'If I'd asked customers what they wanted, they would have told me 'a faster horse!' People don't know what they want until you show it to them. That's why I never rely on market research. Our task is to read things that are not yet on the page." Jobs is right, to an extent. You need to show it to customers, but ultimately customers decide the fate of your products with their dollars. If they don't buy them, it is game over.

Second, keep an open mind and suspend judgment until enough data has been gathered. If the initial data contradicts your initial hypothesis, it could be an anomaly, or it could be a sign that a change is required. Be aware of your own biases and be careful that you are not paying attention only to information that meets your expectations. If the data tracks flawlessly with your opinion, double-check the source to make sure there are no biases built into the process.

When the data contradicts your hypothesis, go back and try to understand why:

- What incorrect assumptions did you build into the hypothesis?
- What did you miss?
- What could you do differently next time so that the same mistake is not repeated?
- Is it possible that the hypothesis is correct, but you have been looking for validation in the wrong places?

"The absence of evidence is not necessarily the evidence of absence; nor is it evidence of presence."

DONALD RUMSFELD

Gathering data and feedback takes time. Be patient with the process. But after you have given the process a chance, be ready to make decisions. Talk to a diverse set of customers and prospects, check your data, give it the time required, remove any biases, and then do a final review before you decide.

Technically this completes the fourth SPX loop, the implementation. All the data generated and gathered thus far and the iterative feedback from all four loops will lead to the final step in the SPX process which we discuss next.

Scale or Terminate Initiatives

In SPX's final stage, organizations will be able to make prompt decisions about each initiative underway by reviewing the results of the feedback. The decisions are made objectively based on data collected throughout the process.
    In general, three choices emerge:

- Scale the initiative
- Terminate the initiative
- Modify the initiative to include other aspects not considered before

Ideally, organizations should decide to either scale the initiative or simply end it. The third option should be used with careful consideration to avoid problems with long-running initiatives that consume organization resources but don't achieve their stated objectives due to indecisiveness or constant tweaking.
    Some organizations choose to conduct formal strategy reviews. In our view, strategy reviews should be an ongoing exercise that is embedded in the entire strategic planning process proposed by SPX. It is not something you schedule once or twice a year. Week in and week out, organizations that want to thrive in the Exponential Era must be reviewing their strategy continuously. Instead of formal reviews, we prefer to lean toward an informal approach to foster a relaxed atmosphere of collaborative discussions. According to Jack Welch, informality is critical to candor. Bossidy and Charan explain: "Formality suppresses dialogue, while informality encourages it. Formal conversations and presentations leave little room for debate. They suggest that everything is scripted and predetermined. Informal dialogue is open. It invites questions, encouraging spontaneity and critical thinking... Informality encourages people to test their thinking, to experiment, and to cross-check. It enables them to take risks among colleagues, bosses, and subordinates. Informality gets the truth out."[119]
    Be careful to avoid long rituals of PowerPoint presentations where there is little or no constructive discussion and almost no decisions. The last thing you want is to spend hours on end reviewing mind-numbing data and never come to any decisive action. The strategy review should not be a forum for maneuvering for power and fighting for budgets. Nor should it be a reprise of the last strategy review. Ideally, this should be an interactive session that generates healthy debates, where all participants have a chance to make their point and speak their minds. This should be an exercise of creativity where ideas are debated, and decisions are made in a continuous flow.

As you conduct your review, several questions should be raised:

- Are we choosing the right ideas?
- Are the ideas synergistic and internally consistent?
- How are the market and the competition reacting?
- Does the feedback validate our assumptions?
- Do we have the capabilities to execute the strategy?
- Are people committed to them?

The discussion about capabilities is particularly applicable here. Remember that in the process of defining priorities we had some horizons that required the development of additional capabilities in order to pursue them. As you review your strategy, ask pertinent questions about where the organization stands in the development of these capabilities:

- Do we have the salesforce to win in the new market segments?
- Do we have enough technology talent, and are we able to keep up with future developments?
- Are the plans too ambitious?
- Are we taking more than we can handle?
- Do we have the capital required to fund the initiatives?

Ultimately this process will result in decisions about what initiatives to scale, modify, or terminate. Once these decisions are made, you will continue progressing through the SPX Flywheel and repeat the entire process. This entire iterative process, once ingrained into the organization's culture, starts to set the stage for intercepting the future and generating sustainable value.

CHAPTER 9

Leading a Culture of Change

The SPX Flywheel with its four loops sits on top of two foundational elements: *Culture and Behaviors* and *Executive Engagement*. In this chapter, we will unpack the leadership characteristics that shape the change mindset and that influence behaviors found in organizational cultures embracing innovative thinking.

We believe it all starts at the top. Without executive engagement, organizations simply don't have a fighting chance of winning in an environment that is changing at exponential speed and that requires constant innovation. Successful organizations in the Exponential Era require a leadership team that is disciplined at investing time and continuous effort into the strategic planning process. As mentioned before, just saying you have a strategy does not make it so. And without a strategy, you don't have a way for the organization to know where they are going and what they need to do to reach strategic goals. Senior leaders need to engage in the constant mapping of opportunities, risks, and capabilities and make decisions that are difficult in nature. The toughest decisions will inevitably find their way to the top.

Senior leaders must exemplify the change-embracing culture that is expected of organizations that thrive in the Exponential Era. CEOs set the tone. If they surround themselves with "yes-men," shoot the messenger, reject signals that contradict their internally focused biases, and tolerate decisions based on long-held beliefs that no longer match reality, they will not be able to lead the organization in an era that is unforgiving of slow-moving bureaucracies. The ego-driven leader-hero days are gone. The effective CEOs highlighted in

The Exponential Era: Strategies to Stay Ahead of the Curve in an Era of Chaotic Changes and Disruptive Forces, First Edition. David Espindola and Michael W. Wright.
© 2021 by The Institute of Electrical and Electronics Engineers, Inc.
Published 2021 by John Wiley & Sons, Inc.

the book would be the first ones to tell you it is all about teams and leaders that can influence behaviors across their organizations.

"If you don't like change, you are going to like irrelevance even less."
GENERAL ERIC SHINSEKI

Unlike leaders in the past, the vast majority of today's leaders cannot simply order things to be done. They must set a vision that is compelling, exciting, and engaging. The organization must understand and embrace its "why," its purpose. The leadership team must persuade the organization to work in unison in the passionate pursuit of a common goal. An engaged organization is one that is empowered to take initiative and make decisions. If the CEO is the single decision-maker in the organization, it will discourage creativity and stifle innovation. Even though tough decisions will often find their way to the top, decentralization encourages leadership at all levels of the organization, resulting in more effective outcomes.

"People can be divided into three groups: those who make things happen, those who watch things happen, and those who wonder what happened."
JOHN W. NEWBERN

Digital technologies have completely changed the competitive landscape. As described earlier, competitive advantage is now achieved by the unique ways in which companies access advanced data capabilities and seamlessly integrate their core competencies into a cooperative ecosystem of partners. This disruption to conventional business models has left many leaders in a state of confusion, paralysis, and sometimes denial. Incumbent decision makers are oftentimes unable to see that the rules of the game have changed and that the traditional premises under which they operated have become obsolete. The problem becomes worse when these leaders invest personal and organizational capital in promoting a point-of-view that is no longer applicable, doubling down on commitments to ideas that have been superseded. The assumption of knowledge where none exists can be very dangerous.

We recognize the difficulty faced by incumbent leaders. There is a real tension between exploiting a currently profitable, well-known, operationally efficient business model and exploring unproven new ones. The problem becomes particularly acute when organizations seek to find repeatable operational solutions to complex problems that require innovative thinking.

Satya Nadella found Microsoft afflicted with this issue when he took the reins. According to his book, *Hit Refresh*, Microsoft struggled with a culture in which to gain credibility and be accepted, employees needed to look smart.[120]

Nadella realized that this attitude was consistent with the fixed mindset described by Carol Dweck in her research. People with a fixed mindset spend a disproportional amount of time proving how good they are, preoccupied with being right. People that have a growth mindset, on the other hand, are inclined to learn from the experience, staying open-minded about new information coming in and being less concerned about looking good and caring more about getting better.[121]

To counteract the problem of being committed to obsolete ideas, a change in leadership may be required, as was the case with Microsoft. Instead of large commitments to a single preconceived idea about the future, what organizations need in the Exponential Era are leaders that embrace a discovery-oriented growth mindset, promoting changes in behavior and encouraging small investments in experimental activities. The emphasis must be on continuous learning through a systematized feedback loop, which, we hope we have made clear, is embedded in every part of the SPX process.

Exploration versus Exploitation – the Innovator's Dilemma

Even a fresh CEO, like Nadella when he took the top leadership position at Microsoft, has to contend with the pervasive issue of finding the right balance between exploration and exploitation. Clayton Christensen brought new light to this problem in his classic and authoritative book titled *The Innovator's Dilemma*, where he describes how a successful company with established products will get pushed aside unless managers know how and when to abandon traditional business practices.[122] This is a very difficult issue for leaders to contend with, especially when leading a large and profitable enterprise.

In searching for additional insights on how to best deal with this dilemma and staying true to our practice of using lateral thinking to solve complex problems, we have explored – yes, definitely pun intended – how other domains have dealt with this fascinating topic. It turns out that computer scientists have been working on finding this balance for over 50 years – they call it the explore/exploit tradeoff. In Computer Science, exploration is analogous to gathering information and exploitation is the use of information already gathered to achieve some result. The tension between exploration and exploitation is described in Computer Science by the "multiarmed bandit problem." The name comes from slot machines, and the question is whether one should keep pulling the arm on the same slot machine or switch to a different one in order to maximize gains. Interestingly, this problem and the insights gained studying it apply to many aspects of human decisions. Mathematician Peter Whittle states that the bandit problem "embodies in essential form a conflict evident in all human action."

But instead of focusing on a fixed time interval – i.e. the number of times you can pull the arm of a slot machine on a single visit to a casino – a mathematician named John Gittins approached the problem with the goal of maximizing payoff for a future that is endless yet discounted. His thinking follows the idea that the value of payoffs decreases the further into the future they are, analogous to how finance uses a discount rate to arrive at current value for future payments. Following this logic, he created a dynamic allocation index that became known as the Gittins index. Applying the Gittins index is fairly involving, but its practical usage can be simplified. For example, if you assume that a payoff next time is worth 90% of a payoff now, and you have no experience with the potential new situation – such as a new slot machine – it is more attractive to pull the arm of the new machine than to continue pulling the arm of a machine you know pays out 70% of the time! With the future weighted more closely to the present, the value of making a chance discovery relative to a sure thing goes up even more. For instance, if the payoff next time is worth 99% of the payoff now, an untested machine has an 86.99% chance of a payout. Again, the context is an endless future, so Gittins is assuming an infinite sequence of pulls, each pull optimally balancing exploration and exploitation in a dynamic setting. As explained by Brian Christian and Tom Griffiths in their book *Algorithms to Live By: The Computer Science of Human Decisions*, "exploration in itself has value, since trying new things increase our chances of finding the best. So, taking the future into account, rather than focusing just on the present, drives us toward novelty."[123]

The Gittins index has its limitations as it is optimal only under strong assumptions, including the discounting of future rewards that value each pull at a constant fraction of the previous one, and the infinite sequence of pulls. It also does not account for the cost of switching. But it does provide a generic framework that encourages exploration, especially when we don't have experience with the subject matter. Another approach to the problem uses the same framework Jeff Bezos used when deciding to start Amazon – the now-famous "regret minimization framework." To find an effective solution to the regret minimization framework, Herbert Robbins and Tze Leung Lai, mathematicians at Columbia University, came up with the Upper Confidence Bound algorithm. It basically says that as we gain more data about something, its "confidence interval" will shrink. It also states that we should pick the option for which the top of the confidence interval is highest. So, the less we know about something the more we should explore. Christian and Griffiths explain it this way: "Like the Gittins index, the Upper Confidence Bound is always greater than the expected value, but by less and less as we gain more experience with a particular option." The principle it operates on has been called "optimist in the face of uncertainty." Christian and Griffiths add that "by focusing on the best an option could be, given the evidence obtained so far, these algorithms give a boost to possibilities we know less about. As a consequence, they naturally inject a dose of exploration into the decision-making process, leaping at new options with enthusiasm because any one of them could be the next big thing."

What we can infer from Computer Science is that if you are in for the long term, and you don't have much experience with the subject matter at hand, you should tilt the balance toward exploration. In the Exponential Era, where discoveries are being made and technologies are converging at unprecedented rates creating unimagined new scenarios, our experience in any particular domain will progressively diminish. So, unless you are in business for the short-term – perhaps getting ready to retire and considering selling a long-standing profitable outfit – science shows that exploration is the best choice when trying to minimize future regrets.

Amplifying Flow Through the Organization

Companies that lead a culture of change encourage people throughout the organization to provide input into its strategy and innovation planning process. For example, Adobe's Chief Strategist and Vice President of Creativity, Mark Randall, created a program called Kickbox which solicits ideas and input from the general employee population. The impetus for the program came from Randall's interview of dozens of employees which revealed a commonly found obstacle to the dissemination of interesting ideas: fighting its way through the bureaucracy to get the approvals necessary. Randall's insight was that for the price of funding just one $1 million project, he could fund one thousand little $1,000 bets. Projects teams that make it through all the levels of the program get a chance to pitch their ideas to key decision makers in the organization.[124]

Rita McGrath emphasizes in her book *Seeing Around Corners* how a strategy that is communicated effectively provides clarity of purpose for the organization. She says she is often asked if the concept of strategy and its long-term implications are of any use at all since competitive advantages are transient and the next big thing is impossible to predict. She responds by reaffirming the importance of communicating a well-thought-out strategy: "In a complex situation, when you want to empower the entire organization to be able to act without direction from the top, having a shared view of what the purpose is and how each participant fits into it is absolutely critical. It is only with a basis of shared understanding of what we're all trying to achieve here that distributed action is possible." She further adds that "a key to navigating successfully through inflection points is the ability for everyone in an organization – not just the leadership – to spot inflection points and to mobilize action to take advantage of them."

Encouraging and setting a psychologically and emotionally safe environment allows the organization to provide strategic input, experiment with ideas, and make the right decisions. Acquiring input across the employee population requires the leadership team to embrace and encourage the free flow of information, allowing these ideas to be born and run through the organization all the way to the decision makers and back in a continuous flow that is as open as possible. Leaders must be open to receiving, dealing with, and providing "air cover" for bad news, as disappointments will inevitably challenge even the strongest proponents of innovation. When Microsoft inadvertently created an AI chatbot that learned to send back messages that were racist, sexist, or with other negative connotations, the team responsible quickly shut it down but might have wondered how the management team would react. Nadella sent the following note to the team: "Keep pushing, and know that I am with you."[125] When people try things that don't work, they need to know that they will be supported by the senior leadership team.

Nadella's attitude of seeking experiences that challenge his worldviews and providing air cover when there are setbacks is an ultimate example of leadership. No one likes to give bad news, but how they are received at the top of the organization will determine whether bad news ever gets to the top at all. A receptive leader will be confronted with bad news. A leader that shoots the messenger will be comforted by the silence that permeates the thin veneer of his ignorance.

Mobilizing the Organization Through an Inflection Point

According to Jeff Bezos, the biggest challenge in dealing with an inflection point is understanding and accepting the implications of the traditional way of doing business, deciding what opportunities to pursue, and navigating the organization through it. Mobilizing the organization through an inflection point can be difficult because most people have never been through one before, and most are rewarded to keep doing what has always worked in the past.

Leaders need to empower the organization with the courage and the methodologies to navigate through an inflection point. The senior leadership team must get behind the organizational effort to develop new capabilities and learn fast even in a time of chaos and uncertainty. Leaders must also assure that there are team alignment and prioritization. In times of complex and fast changes, internal infighting only adds to the chaos and is detrimental to providing clarity of purpose and direction to the organization. All team members need to hear directly from senior leaders, and particularly from the

CEO, a message that reinforces the company's purpose, mission, and priorities. Communication is of utmost importance.

Rita McGrath's research suggests that "given the amount of disruption in the business environment" characteristic of inflection points, "a core strategy is more important now than ever before, as without real clarity on strategy and priorities, the crescive approach will disintegrate into rudderless activity." She also references the research of her colleague Thomas Kolditz, a retired brigadier general who built much of the leadership curriculum at West Point. Kolditz's research suggests that leaders in a precarious situation need to keep people calm and "establish the vision to the way ahead, even if there is no detail to it." Providing a sense of shared risk is also critical. People want to follow a leader that can relate to their issues, who share a common mindset, and who do not set themselves apart or insist on special perks.

Leaders must carefully select the people that will carry out the key priorities of the strategic plan. The very best people must be placed on the most promising projects. Lou Gerstner once said that great companies put their best people on opportunities, average companies put their best people on problems.

However, just selecting the best people is not sufficient. It is also critical to put them in teams that can work effectively together. To do this, leaders must understand how people are connected to each other in the organization – not from an organizational chart standpoint but from a relationship point of view. Success requires trust in the networks across organizational siloes. It starts with who knows whom, and who trusts whom. Here is how Nadella describes building a team-oriented culture at Microsoft: "I've optimized for people who want to work as part of a team. . . To run the show, you have to work as a team."

Highly effective teams are built on trust among team members and with the leader. When a team operates at optimum effectiveness, there is little, or no politics involved. No one criticizes anyone behind their backs. Ideas are debated in a constructive, open manner. Each team member understands their strengths and weaknesses and does not get defensive if a suggestion or comment is challenged. In such an environment, team members support each other on all fronts, intellectually, emotionally, and psychologically, and never let outside forces break the trust that exists within the team.

Knowing When to Quit

A culture of change embraces a mindset that gives permission to quit. It recognizes that the judicious decision to no longer pursue a previous choice, decision, or initiative that is no longer meeting expectations is a competitive advantage. Resistance to change is sometimes reinforced through well-intentioned, but not necessarily effective clichés such as "never give up." Steven Levitt, the author of *Freakonomics*, conducted experiments to understand people's willingness to make life changes, such as switching jobs. He found that those that followed a random suggestion to change jobs were substantially happier six months later than those that did not follow the suggestion. According to Levitt, the study suggested that "admonitions such as 'winners never quit and quitters never win,' while well-meaning, may actually be extremely poor advice."[126]

Seth Godin discusses the importance of knowing when to quit in his book *The Dip: A Little Book That Teaches You When to Quit (and When to Stick)*. The book criticizes the idea that "quitters never win." He makes the case that people that reach the apex of their domains quit fast and often as they navigate through a discovery process searching for a best-fit. According to Godin "we fail when we stick to tasks we don't have the guts to quit."[127]

Godin does not advocate quitting just because the pursuit is difficult. Psychologist Angela Duckworth conducted several studies across many fields, from West Point cadets to spelling bee contestants, and showed the importance of what she calls "grit." Her book, *Grit: The Power of Passions and Perseverance*, has gained much acclaim. But Duckworth herself warns against judging people that have made the choice to quit: "I worry I've contributed, inadvertently, to an idea I vigorously oppose: high-stakes character assessment."[128] Godin recognizes the importance of persevering through difficulty but recommends that people consider the conditions under which they would quit before they engage. He makes the point that one must make the distinction between a failure of perseverance and thoughtful recognition that better options are available.

Godin also notes another condition that contributes to resistance to change, and that leads to misguided decisions: the "sunk cost fallacy." Humans have a difficult time leaving something that we have invested time and money in, even if it is no longer benefiting us. In the book *The Confidence Game: Why We Fall for It. . .Every Time*, Maria Konnikova demonstrates that the sunk cost fallacy is so deeply entrenched, that it is used by astute conmen that understand the psychology behind it to get you hooked on making bad decisions you wouldn't make otherwise. They start by asking small favors and investments, and once you have made the initial commitment, it becomes a lot more difficult to say no to bigger asks. Once an investment in time, energy, or money has been made, instead of walking away from the sunk costs, people tend to continue investing beyond any rational explanation. "The more we have invested and even lost, the longer we will persist in insisting it will all work out," she wrote.[129]

Knowing when to quit is a subject of interest that goes beyond the humanities. Computer scientists have also grappled with this question and have developed algorithms to determine what they refer to as "optimal stopping." Christian and Griffiths describe the 37% Rule, an optimal stopping answer to the question of how many options should one consider, derived from the famous puzzle known as the "secretary problem." The puzzle can be summarized as follows: you are searching for a secretary, so you go through an interview process intending to find the best applicant. The question is, when do you stop searching? Taking the first best applicant seems rash. If you have a pool of 100 candidates, interviewing all of them despite having found great candidates mid-process seems inefficient.

It turns out that the optimal solution is provided by what Christian and Griffith refer to as the Look-Then-Leap Rule. The rule says that you search for a predetermined amount of time or number of candidates to explore potential options – the "look" phase. During this time, you don't commit to anyone, no matter how good they are. Then you enter the "leap" phase where you commit to anyone who outranks any candidates from the look phase. So how do you know when to transition from the look to the leap phase? Scientists and mathematicians have demonstrated that the optimal time is 37% – thus the name 37% Rule. So, after you have interviewed 37% of the candidates in the pool, choosing none, be ready to leap and hire the first one that is better than anyone interviewed so far. So, if you are looking for a science-based rule of thumb for knowing when to quit, spend time searching through about 1/3 of possible choices without making any commitments and then be ready to choose the best choice from that point forward.

Sticking to a strategy that no longer serves the organization is a sure way to be left behind in the Exponential Era. Seth Godin has given us some powerful insights about gaining competitive advantage by knowing when to quit, and Computer Science provides additional discernments through algorithms that can be used as frameworks for making decisions about when to quit.

Executive Engagement

We started this chapter explaining that Executive Engagement is a foundational component of SPX. But what exactly does it mean for executives to be "engaged"? Simply stated, it means being where the action is. In previous chapters, we described the disastrous consequences of CEOs and management teams who were out of touch with the day-to-day realities of their companies. To be engaged, leaders must live their business by getting as much information as possible directly from the source. They can't be satisfied with having information delivered to them by direct reports or staff people who filter the information and add their own perceptions, biases, and agendas.

In practical terms, this means visiting with stakeholders, both internally and externally, whether they are in stores, manufacturing plants, or offices spread throughout the globe. They must meet with local leaders, see what kind of teams they have, and encourage an open dialogue. This will give the senior leaders a chance to connect personally with these people and get an intuitive feel for the way they interact, communicate, and operate.

This is particularly critical when starting a new initiative. Many initiatives start with enthusiasm and fanfare but fail to get traction and end up dead after a few months. The reason is that deep in the organization managers despise additional time-consuming activities that distract them from their day-to-day operational responsibilities. The leader's personal involvement and commitment will help overcome resistance. The key is to engage at a much deeper level than just announcing the initiative – the leader needs to understand how the initiative will work and clearly communicate the benefits.

Once initiatives have been launched, and goals and priorities have been clearly communicated across the organization, leaders must follow through to make sure people are made accountable for results, and that things actually get done in the right way and through acceptable behaviors that align with the organization's values. It is critical to put the right metrics in place and to monitor progress as the initiatives progress.

Another critical aspect of executive engagement is coaching. This is perhaps one of the most important aspects of a leader's job, passing on to the next generation of leaders all the knowledge and experience accumulated over an entire career and influencing the next generation's behaviors. Coaching expands the organizational capabilities, and when done correctly, builds trust and rapport between leaders and their subordinates.

An effective coach will know how to ask questions that will help surface the help people need. Formal classroom education is an important component of expanding organizational capabilities, but many organizations miss an opportunity to zoom in on the specific needs of individuals or the organization at large by indiscriminately sending people to generic management and leadership courses. These courses, like any other tool, are only effective if put into practice. This needs to be done thoughtfully to make sure it is addressing a specific company need, as part of a larger coaching effort, where every leader is a teacher.

Executive engagement requires a conscious effort by the leadership team. It requires leaders who are comfortable with their own strengths and weaknesses and who are willing to learn from their mistakes and successes. No leader can be good at everything all of the time, from developing strategies to managing people, to running efficient operations, to closing deals.

The best leaders know their shortcomings and surround themselves with others who can provide complementary skills. They engage the entire team to help each other, and through daily engagement, live the values of the organization. Through the passionate pursuit of purpose, coupled with tenacity, they influence the behaviors of the organization, establishing the organizational culture. "Performance is the residual of behaviors – influencing behavior is the essence of leadership."[130]

Culture and Behaviors

The second foundational component of SPX is culture and behaviors. Again, let's start with some definitions. What exactly are culture and behaviors? Culture is the set of values, beliefs, and norms of behaviors that guide the social interactions in an organization. Behaviors are the specific actions that people take, and how those actions are conducted, as they perform their duties and run the business. It includes how people behave in meetings, how they communicate formally and informally, how they build relationships, how ideas are shared, and how learning occurs. These behaviors, linked to values, beliefs, measurements, and reward systems will ultimately determine how the company performs.

The analogy of a computer system, where culture and behaviors are the software whereas strategy and organizational structure are the hardware has often been used to illustrate this interoperable synergy. One cannot function without the other. But it is the software, the management "soft stuff," that integrates the organization into a synchronized, unified whole.

A challenge that many companies face is the inability to set specific actions required for cultural change. It is easy to say: "we need to change our culture," but it is a lot more difficult to be specific about the "how." Hiring a consulting firm to do an employee survey and present a cultural diagnostic is not sufficient. A discussion on cultural change needs to get to the specifics of "from what we are to what we will become." Once that is understood, the next question is where to start. The simple answer is that it needs to start with the leadership team. The leaders of the organization must exemplify the types of behaviors that they expect from their organization. Anything short of that will quickly be dismissed as just another "slogan of the month." Leaders need to know "what you do does" – what their leadership behaviors and decisions do to influence others and the resulting outcomes.

Similar to how designers build their way into a good design, instead of just thinking their way through it, leaders need to act themselves into a new way of thinking in order to change the culture. That is where behaviors come into play. Behaviors are beliefs turned into action, resulting in performance. Behaviors are shaped by a set of norms, or "rules of engagement" that determine what is acceptable and what is expected from individuals and teams. To influence behaviors, leaders must align measurements and rewards with the expected behaviors and make those alignments transparent. This is what will tell the organization what is valued and recognized by the culture as it moves from today to tomorrow.

The behaviors that the organization will exhibit is a reflection of its leaders. This cannot be emphasized enough. A leader who is not engaged in the daily life of the business cannot influence or change its culture. As one of the authors wrote in *The New Business Normal*: "Leaders are like eyes, a window into the soul of the organization."[131]

One company that has done an outstanding job in articulating their values and the expected behaviors of their existing and prospective employees is Netflix. Netflix truly provides one of the best illustrations we have seen. They have posted their expectations on their job website for all to see, and we would highly encourage you to go to their website and read the entire list. Here we have selected a portion to exemplify it: [132]

Judgment

- You make wise decisions despite ambiguity
- You identify root causes and get beyond treating symptoms
- You think strategically, and can articulate what you are, and are not, trying to do
- You are good at using data to inform your intuition
- You make decisions based on the long term, not near term

Curiosity

- You learn rapidly and eagerly
- You contribute effectively outside of your specialty
- You make connections that others miss
- You seek to understand our members around the world, and how we entertain them
- You seek alternate perspectives

Innovation

- You create new ideas that prove useful
- You reconceptualize issues to discover solutions to hard problems
- You challenge prevailing assumptions and suggest better approaches
- You keep us nimble by minimizing complexity and finding time to simplify
- You thrive on change

Integrity

- You are known for candor, authenticity, transparency, and being nonpolitical
- You only say things about fellow employees that you say to their face
- You admit mistakes freely and openly
- You treat people with respect regardless of their status or disagreement with you

Netflix has dedicated a considerable amount of time and effort to get their culture right. They are known for clearly communicating and living these values and expected behaviors in their day-to-day business conduct. The linkage between behaviors and performance is clear – the phenomenal results Netflix has been able to achieve speak for themselves.

SECTION THREE

CHAPTER 10

The Exponential Human – Social and Ethical Challenges in a Chaotic World

At the beginning of 2020, the world started hearing about a new virus that was starting to spread in China. The outbreak was first documented in Wuhan, in the Hubei Province. The initial information being divulged out of China indicated that this virus was similar to the SARS-CoV and MERS-CoV viruses that had mostly impacted Asia and the Middle East in the last two decades. The signs of infection were similar to the common cold and included respiratory symptoms like dry cough, fever, and shortness of breath. This virus, which was part of the coronavirus family, was called SARS-CoV-2, and the disease labeled COVID-19. Despite the similarities to its predecessors, this virus had one characteristic that made it unique: it spread exponentially.

SARS cases totaled 8,098 in 17 countries, resulting in 744 deaths. MERS cases totaled 2,500 in 21 countries, resulting in approximately 860 deaths. On March 11, 2020, the World Health Organization (WHO) declared the COVID-19 a pandemic. In March 2020 there were over 1 million confirmed cases and over 100,000 deaths across the world.[133] Towards the end of 2020 cases surpassed 45 million, and deaths more than one million.

What is remarkable about COVID-19 is how quickly it changed the world. No one was ready for this. Hospitals were ill-equipped to deal with the number of Intensive Care Unit (ICU) beds required. The lack of ventilators forced health providers in parts of the world to make extremely difficult ethical decisions, having to choose who would live and who would die. Protective gear was lacking, putting front line workers at risk. Suddenly, store shelves

The Exponential Era: Strategies to Stay Ahead of the Curve in an Era of Chaotic Changes and Disruptive Forces, First Edition. David Espindola and Michael W. Wright.
© 2021 by The Institute of Electrical and Electronics Engineers, Inc.
Published 2021 by John Wiley & Sons, Inc.

were emptied as people stockpiled, bracing for disruptions in the supply chain. Shelter-in-place orders forced people to stay home, causing economic mayhem. People started losing their jobs, and the US government authorized an unprecedented relief package of two trillion dollars. All in the span of a few weeks.

We could not have thought of a better example to illustrate the chaotic and sudden nature of exponential change and its social, economic, and ethical implications. Many of the changes we are experiencing now, such as online learning, joblessness, scientific acceleration in search for cures, innovative business models, decisions that change from one moment to the next based on new information, difficult ethical and economic choices, etc., are the types of challenges that we should expect in the Exponential Era. COVID-19 is just a catalyst for the inevitable. A wake-up call. A dress rehearsal for the profound and difficult choices we all have to make in a world of chaotic changes and disruptive forces.

Let's be clear that we are not predicting doom and gloom. As we saw in Chapter 2, the megatrends of the Exponential Era will bring numerous benefits to humanity, including longer and healthier lives, general abundance, better transportation, more awareness of and harmony with the environment, and a noticeable reduction in mundane and physical labor. But it does come with a volume and velocity that we are not accustomed to, making it difficult to adapt. It forces uncomfortable changes. It causes social, political, and ethical challenges – just like COVID-19 has and future pandemics will continue to do. We will examine some of these challenges in this chapter.

The Workless Society

The economist Daniel Susskind in his book *A World Without Work – Technology, Automation and How We Should Respond*, tells a compelling story that illustrates the threat of technological unemployment. His analysis validates our belief that robots, AI, and automation in general will encroach all aspects of human work, including manual, cognitive, and creative domains.

Susskind shows that historically we have been able to subside the fear of technological unemployment because the complementing force of technology has always been able to create more job opportunities than the substituting force that results in the elimination of existing jobs. However, evidence shows that this time is truly different. If you just look at the trend of the number of employees required to run the most prominent US companies this becomes abundantly clear. In 1964, AT&T had 758,611 employees and revenues of 23.62 billion in 2020 dollars, a rate of $31,135 per employee. In 2018, Apple ran its business with a fraction of that number of employees, 132,000, and posted $265.6 billion in revenues, a rate of $2,021,121 per employee. This is a staggering 65× improvement in sales per employee in just about half a century.

The impact that digital business models are having on the reduction of the number of employees required by emerging companies is astounding. YouTube had only 65 employees when it was acquired for $1.65 billion by Google. Instagram had 13 employees at $1 billion acquisition value and WhatsApp 55 employees at $19 billion when acquired by Facebook. Furthermore, research shows that new industries created in the twenty-first century, the majority of which are digital, account for just 0.5% of all US employment. This clearly illustrates the role that a digital business model has in slowly, then suddenly, increasing productivity and reducing dependence on human labor.

Based on this irrefutable evidence, Susskind predicts the dawn of the workless society: "As time goes on, machines continue to become more capable, taking on tasks that once fell to human beings. The harmful substituting force displaces workers in a familiar way. For a time, the helpful complementing force continues to raise the demand for those displaced workers elsewhere. But as tasks encroachment goes on, and more and more tasks fall to machines, that helpful force is weakened as well."[134]

In the short term, as with other historical transitions, we will face a mismatch of skills where people in the middle who have access to the needed training and support, will have a choice of obtaining the skills to move up, or otherwise move down the skills ladder. Moving up will become increasingly more difficult at all levels of work. Moving to lower-skilled jobs will create a mismatch of identity where people will prefer unemployment rather than moving into substandard or mundane, lower-paid, lower-skills work. In the long term, structural technological unemployment will challenge our notion of work as well as our existing social, economic, and political systems.[135]

A workless society is one of the most difficult challenges for humanity to grapple with in the Exponential Era. What will life be like without work? What will we do in a world in which robots and AI take care of most of our material needs? What purpose will we find in life?

COVID-19 is a leading indicator that may help us get a preview of the answers. It provides early signals of what that future life might look like. We don't mean to dismiss the suffering from sickness and from the loss of loved ones, nor the economic hardship that this terrible disease has brought to the world. But imagine for a moment a fictitious COVID-19 world without the sickness, the losses, and the economic hardships, that potentially represent our future workless reality.

For many of us, the crisis forced us to slow down. It gave us new perspectives on what is really important. It afforded us more time to spend with our families, to sit at the dinner table together, to have more meaningful conversations. It allowed us to connect with others, albeit remotely, in ways that we had not done in a long time. We are paying more attention to our elderly parents, calling them more often, checking on them, and just enjoying the time we still have with them, listening to their stories, absorbing the wisdom that was built over an entire lifetime.

We find ourselves sitting in front of a fire, talking, telling stories, appreciating the company of loved ones. A glance at the night sky reveals discoveries that we were unable to see before. The sky seems so much clearer – possibly as a result of less pollution – and the stars so much brighter. Tiny bright moving objects that we had never seen before – perhaps more satellites that are starting to orbit the earth – catch our attention. We take more walks together, listening to the birds sing. Maybe alternatives to the traditional are not so bad after all.

Instead of the hustle and bustle of the modern world, suddenly, we became more introspective. Why do we work? If shelter, security, food, and health are accessible to the majority of people on the planet – and that is a big if – is there anything else that we really need, or are we just wasting precious moments, destroying the planet, missing the best that life has to offer?

If the abundance promises of the Exponential Era come to fruition in the next few decades, and we address the big "if" raised above, what will it mean to be human and what kind of people will we become? Will we turn into a more introspective connected species that care for one another and focus on the collective bigger picture? Or will we turn into zombies, bored, addicted, grabbing on to vices in search of meaning and purpose?

These are the types of questions our society will be faced with as we enter an unprecedented moment in human history where work is potentially no longer a necessity, but something many will be able to do for the simple joy of creating, imagining, collaborating with others, and being part of a greater collective good.

Next, we present a few areas of opportunities and challenges that require innovative thinking and ethical considerations as we adapt to this new era.

Education Reinvented

Our work on the advisory board of the Technological Leadership Institute (TLI) at the University of Minnesota helped us gain some insights into the challenges faced by our educational institutions. Out of the many discussions that we have had and the numerous recommendations that the advisory board made to the leadership team at TLI, two stood out. The first was the need to put more emphasis on ethics in the curriculum. We could foresee how the ethical challenges faced in the Exponential Era, which we will discuss in more detail later in the chapter, would become a critical discipline and learning priority for future technology leaders. TLI took our recommendation to heart. They added ethical components to many of their classes, and greatly expanded the scope of ethical discussions throughout the curriculum.

The second recommendation was to embrace online learning and distance delivery. It was obvious to all of us that traditional institutions like universities were going to be disrupted by a digital operating model, and we urged them to get started immediately. The world was changing all around them, the temperature was rising very rapidly, and we did not want to see them turn into Flash Boiled Frogs. They listened. They agreed with our points. But there was no significant progress. Until COVID-19 hit.

Suddenly, there was no choice. In a matter of weeks, the entire university had to move to an online, distance delivery mode. Something that found resistance for so long could no longer be deferred. Professors scrambled. Some struggled with the new technology and delivery mode. But adaptation came faster than anyone could have predicted. Delivering a quality education online was not so difficult after all. While there will never be a substitute for learning in groups or the need for socialization that comes with attending school, there is no doubt the frequency of in-person interaction will change and the venues will morph to accommodate the new reality.

Now students at many universities are questioning if they will get some tuition money back. A student shared her thoughts with MarketWatch: "I believe we should, but it hasn't been talked about yet!" Robert Kelchen, a Seton Hall University professor who studies financial access to higher education explained to MarketWatch that "a lot of colleges simply can't afford to give [tuition] refunds. They don't have the extra money to do that when they are still paying their employees."[136]

If a quality education can be provided online, and professors can deliver their content much more efficiently through digital channels, why do students have to continue accumulating life-altering debt and pay tens of thousands of dollars in tuition, fees, and room and board? These existential questions for traditional education institutions are starting to surface at exponential speed. An opportunity suddenly became a threat in just a few weeks. Again, COVID-19 is just a catalyst for the inevitable. In the Exponential Era, reality changes at unimaginable speeds, driven by many different types of catalysts, some technological, and some as a result of a black swan that no one could foresee coming.

The resistance to change we see in academic institutions has a long history. Education was supposed to be disrupted by now, but surprisingly, the credentialing problem, which is the difficulty in proving what one knows, has been a difficult barrier to overcome. With the lack of better tools, many employers still use the degree as a filtering mechanism that makes it easier to deal with the volume of job applicants. This reinforces the problem as students have no choice but to pursue the credentials provided by a college degree in order to access opportunities.

However, this is starting to change as organizations start to experiment with badges and certifications that offer the possibility of pursuing lifetime learning. This alternative to credentialing and knowledge acquisition on demand will be much more sensible in a world where your knowledge is becoming obsolete in the blink of an eye.

The economics of higher education only functions because there are no alternatives. Universities are in business only because of the all-commanding credentialing of a college degree. But as alternative credentialing mechanisms emerge, and the efficiencies afforded by online education become apparent, the business model of the university crumbles. Clayton Christensen predicted that the "brands" of universities would be replaced by the brands of individual superstar professors.[137] Imagine a world where you have multiple badges obtained through a lifetime of study, certified by the best professors in the world through online interactions. How much value would that have compared to a four-year degree from Harvard?

What will the future of education look like? Research provides some early signals. Sugata Mitra, professor of education technology at Newcastle University, showed that illiterate children in slums in India were able to learn how to surf the web and teach themselves the basics of reading and writing by simply accessing a computer connected to the internet. Other experiments showed that children can unlock self-directed learning and become technically sophisticated if given the right equipment. Kitkit School in South Korea and Onebillion in Kenya created software that, in an hour a day, produced the education equivalent to what Tanzanian children would have received by attending school on a full-time basis. Based on these findings, and given the low cost of connected devices, we believe that children across the world will have a better chance of gaining access to the education they need in the near future.

With the convergence of virtual reality, AI, and 5G, an immersive high-quality education can be provided at affordable cost, and in some cases even free. According to Peter Diamandis, a medical student, for instance, will be able to do a virtual autopsy and strip away layers of skin and muscle in a navigable operating room. 3D printing is already producing anatomically correct organs, layer by layer, designed to help surgeons plan and practice complex procedures and train new surgeons on organs that respond as a patient would. Students will have much richer learning experiences in a 3D environment, whether virtual or physical, making it much more likely that they will retain their learnings in long-term memory.

In addition, the convergence of AI and IoT sensors will allow the development and delivery of customized learning environments. Neurophysiological data and tools will allow AI to put students in a growth mindset, or get in a state of flow, which research has shown can greatly enhance learning.[138]

One thing is certain. The reinvention of education is long overdue and offers tremendous opportunities for innovators. As we face the challenges brought by technological unemployment, we need to be able to very quickly retool and retrain people capable of acquiring the increasingly difficult skills that will be in demand in the twenty-first century. We hope that through the convergence of technologies, we can educate vastly more people in shorter time periods and in much more effective ways. Perhaps we will be able to alleviate or delay the structural technological unemployment that we discussed earlier, buying additional time to adapt and figure out the social, economic, and political frameworks that will guide the behaviors of society in the future.

Global Synchronous Communications

The explosion of global telecommunications has taken place in the space of two short and fast-moving decades that have seen the pervasive adoption of cellular networks combined with the near-ubiquitous availability of connected devices that are redefining our business and personal interactions in every corner of our world. The convergence of these technologies has enabled virtual and nearly instant connectivity between individuals, families, groups, organizations, companies, governments, teachers, and students in the form of instant messaging, conference calls, video conferencing, and distance learning. We see this as a point of departure, that is, a new horizon.

Most definitions of synchronicity in communications in the past considered only person-to-person or finite and measurable groups. In the Exponential Era, the only limitation to synchronous global communication is network load and time delays measured by the distance electrons have to travel point to point at nearly the speed of light. The number of interconnected networks will multiply and become more cost-effective as satellites are deployed, the flexible cloud becomes ever more accessible, and chips utilizing light reduce the time and cost for transmission.

Our global telecommunications infrastructure and our network of networks have created an inherent transmission capability for data, voice, and video to be accessed, aggregated, analyzed, and acted on, anywhere, by anyone, at any time. Those with greater access to this powerful combination will quickly gain sustainable advantage and possess the potential to reach the Point of No Return relative to competitors.[139]

This globally significant change is one that not only enables the Exponential Era but also accelerates the pace of development, the rate of adoption, the acceleration of collaboration and discovery, and the expansion of knowledge in near real-time. It also provides the backbone for AI to AI global deployment with astounding ramifications. The interconnectivity of AI networks that are driven, managed, and designed by other AIs that are able to communicate in real-time globally with people through brain-computer interfaces will redefine what it means to imagine and collaborate, possibly arriving at what Kurzweil calls the "Singularity" – a single unified collection of all the world's intelligence. It will also challenge our values and systems as the same interconnectivity can be used for mind control at scale, intentional addiction, and weaponized purposes.

Today we have algorithms capable of identifying our moods, passions, thought processes, mindsets, and, yes, our behaviors. For example, the company Realeyes, founded in 2007 by Mihkel Jäätma, is developing what he calls "emotion AI." Realeyes believes that it won't be long before emotion AI is a feature of almost every area of our lives.

"AI is going to be the next industrial revolution," Jäätma predicts. "So it's going to have a huge impact on all industries. And emotion AI is going to be a core part of all AI, so we want to be at the forefront of when that happens and to have the right impact on that as well."[140] Just because technology at the moment is unevenly distributed doesn't mean it isn't here, it just means you don't have access yet. But in the Exponential Era, access to technology will accelerate and surprise everyone, just as the coronavirus – which started slowly and is finishing dramatically – has.

If the algorithms developed by AI can intrude on our being and change our decision-making processes, at scale and velocity, we have entered a new form of existence. The real possibility of mind control is not that far away. According to a new study published in Nature Machine Intelligence journal, the study authors were able to use a "Fully portable and wireless universal brain–machine interface to enable flexible scalp electronics and deep learning algorithms to control an electric wheelchair, a motorized vehicle, and a keyboard-less presentation."[141] The question raised in the brain–computer interface (BCI) world is what happens when the BCI reverses direction and the computer controls the brain? Woon-Hong Yeo, an assistant professor at Georgia Tech, one of the authors of a new study on BCI, stated that "This work reports fundamental strategies to design an ergonomic, portable EEG [electroencephalography] system for a broad range of assistive devices, smart home systems, and neuro-gaming interfaces."[142] Think millions of gamers interacting globally, and suddenly the computer decides to control them. It would be great if we could say this is just science fiction and it's going to be used primarily for aiding humans to more efficiently control things they need and enjoy doing, but that likely won't be true.

BCI research and development is happening across the globe as private, academic, governmental researchers and militaries look to develop brain–computer–brain interoperability. There are numerous participants in this space including Neuralink, Facebook, Alphabet (Google), the US Department of Defense, CIA, and many more. The speed at which these technologies are developed and deployed will challenge our belief systems and our approach to the future.

In the Exponential Era, these belief systems can move at alarming speeds through global synchronous communications. What one day is deemed and accepted as "normal" can change in ever shorter and shorter time horizons. The effects of global communication synchronicity are just being realized and, like all technological transformations, come with both benefits and potential detriments.

Great things can be accomplished with this exciting new level of human–machine global synchronous connectivity. Hopefully the court of time, wherein all truths are eventually exposed, will begin to move as swiftly to root out and silence the bad and harmful as we simultaneously create new ethics for a new era.

New Ethics for a New Era

Too often when social, economic, and governmental constructs change rapidly we find ourselves challenged with the newly created "grey areas" of behaviors. These grey areas are often described as "not quite in focus yet" or "waiting for things to settle out and become clearer." The problem is that at exponential rates of adoption, the ramifications of our ethics are likely to be unforeseen. As Margaret Wheatley so insightfully observed in an essay written nearly two decades ago: "The growing complexity of our times makes certainty about any move or any position much more precarious. And in this networked world where information moves at the speed of light and 'truth' mutates before our eyes, certainty changes and speeds off at equivalent velocity."[143]

In the Exponential Era, we are living with dramatic changes to the space and time relationships of the interactions which define our ethics, and which are the very foundations of our moral codes. What we define as "acceptable" and "moral" codes are behaviors that over thousands of years have developed based on the immediacy of both the co-location of the participants and the actions they take being observed. The diversity of available actions was limited, the predicted range of outcomes was limited, the time frame in which the behaviors were observable was limited, and the context was almost always limited.[144]

Many of those limitations no longer apply. Converging technologies are creating entirely new ecosystems that have never existed before, amplifying the consequences of our actions to much larger scales. In this new era, an action that only takes a few seconds, like the click of a button, can have significant global ramifications. Other cumulative actions over an extended period of time, like the release of carbon dioxide into the atmosphere, result in unforeseen ramifications that will be felt for centuries to come.[145]

Today, leaders must make decisions about adopting and deploying technologies whose impact and ramifications on their people, community, and planet are unforeseeable. Due to the complexity and high adoption rates of modern technologies, decision making has become fraught with overwhelming and interrelated challenges. In fact, the consequences of our technology decisions are no longer obvious. We actively engage in blindly accelerating technologies from innovation to ubiquitous adoption, but we have been unable to foresee the results of our actions over time.[146]

Current examples of unintended consequences and ethical impacts of previous science and technology choices include our energy generation sources affecting climate change, mundane necessities creating oceans of plastic, and the unfettered use of communications subsequently exploding tribalism and enabling the weaponization of social media. Today, the use of virtual assistant devices like Alexa, elucidate potentially irreversible consequences to the choices we are making, as we trade our privacy for the mere benefit of convenience.

Surveillance cameras everywhere may give us a greater sense of security. But at what point do they start to infringe on our liberties, on our ability to freely move about in our neighborhoods without big brother watching? What happens when our most private data can be easily searched on the internet by anyone? Worse yet, what recourse do we have when the data published about us is simply wrong but has been widely disseminated without any trace of its original source? Who owns it? Who is responsible for it? The European Union has taken some steps in establishing policies that try to address some of these data privacy concerns through the General Data Protection Regulation (GDPR). Other regions across the globe are following suit. The challenge is that policies take a long time to be put in place, and in an era where technology moves at vast speeds, by the time the policy is put in place the damage may have already been done.

As the consequences of individual and organizational decisions accumulate, personal responsibility has also grown, and the demand for accountability is rapidly increasing. This threatens the legacy and the significance of leaders everywhere. It also poses numerous economic, ethical, and policy challenges for us all, particularly to those entrusted with the stewardship of technology enterprises, including top executives, investors, and governing boards. Lead organizations, flagship corporations, institutions, and individual leaders will be held accountable not only for what they decided to do but also for what they didn't do, not fully exploring the range of possible decisions and outcomes.

In retrospect, most of the advances in technology to date have been used to address specific problems in closed systems, without regard to dynamic interrelated systems and unintended consequences. What has become apparent is that new technologies and the policies that accompany them can have unexpected results. Therefore, the ensuing questions that we must seek answers for include: What timeframe are we to hold leaders accountable for? What range of constituents, from stakeholders to communities, to future generations, are in the scope of responsibilities of a leader's position? Are we willing to give up some of our well-being today for the future well-being of the yet to be born strangers of tomorrow?

While it will not diminish the responsibility leaders will continue to have for outcomes, it is our hope that the SPX methodology will improve decision-making processes and shed timely and actionable light on difficult decisions, as leaders integrate new ethics into their organizations' behaviors in a time of rapid and chaotic changes.

For My Children

From generation to generation stories have preserved our legacies and history. The values and stories families pass along, fulfill the basic human desire for continuity and meaning. This is how the chain of life is forged. How can we preserve the life stories, family histories, experiences, and knowledge that can provide children with the greatest treasure of all, their family's life stories?[147] The authors have pondered this question for many years.

How do you want to be remembered? Who will help tell your story and the stories of all who came before? What means the most to you that you want to preserve and pass along? What are the things you took a stand for? What has been learned through your own lifetime that could help your children throughout theirs?[148] These are the questions we hoped would be answered and preserved for many generations to come.

In a previous venture, called For My Children, we worked on a technological solution in which we envisioned the personal delivery of such stories. Since then much has changed. It is now possible to create a holograph and to deliver such personal stories in much more vivid ways than we could ever have imagined possible just a few years ago. And in the future, it is likely that even more "magical" means will be available to connect with future generations.

In a fast-changing world, a sense of permanence, of history, of connection through multiple generations become even more meaningful and impactful. We would encourage you to take some time to think about what kind of world you want to leave for your children. Take a moment to record and preserve the memories, the values, and the stories that you want to endure beyond your lifetime.

In the Exponential Era, everything moves faster, including life. Cherish each moment, do greater things, and make it all worthwhile.

A DISCUSSION ABOUT THE USE OF THE TERM EXPONENTIAL

The authors would like to acknowledge that the use of the term exponential and illustrations involving the exponential curve can be a point of contention for those that are more technically inclined. It is not our intent to define the mathematical characteristics of the curve but to simply draw an analogy to the fast and sudden changes that organizations can expect in the current era.

According to Wikipedia, "Exponential growth is a specific way that a quantity may increase over time. It occurs when the instantaneous rate of change (that is, the derivative) of a quantity with respect to time is proportional to the quantity itself. Described as a function, a quantity undergoing exponential growth is an exponential function of time, that is, the variable representing time is the exponent." Merriam-Webster dictionary adds: ". . .approximately expressible by an exponential function. . . characterized by or being an extremely rapid increase (as in size or extent)." We do not believe that the new-fangled and approximate dictionary definition of exponential will blunt the precision of the strict mathematical sense. There are plenty of words in English which hold a precise application in one field, and a broader range of meanings in general use.

We are not making any inferences regarding the starting point nor the duration of the exponential growth for any particular technology or business change. We are simply stating that, in the current era, it is difficult to see the dramatic growth that lies ahead.

We offer an approach that helps organizations prepare for such growth. We argue that once the growth is obvious to all players in the marketplace and a position of leadership has been established, it is very difficult to overcome or challenge such leadership. This is all that we meant by the "Point

The Exponential Era: Strategies to Stay Ahead of the Curve in an Era of Chaotic Changes and Disruptive Forces, First Edition. David Espindola and Michael W. Wright.
© 2021 by The Institute of Electrical and Electronics Engineers, Inc.
Published 2021 by John Wiley & Sons, Inc.

of No Return." It is a business analogy and not a technical definition of the inflection on a curve.

We are also not concerned about the exact technical determination of when an inflection point is observable in any particular time interval. Our concern is to help organizations detect signals that indicate a potential shift in business risks and opportunities and take action before it is too late. In this sense, an exponential curve is simply a visualization tool, and the word exponential is often used in the book colloquially to signify fast changes.

NOTES AND REFERENCES

Introduction

1. Silverthorne, S. (2018). "Amazoned: Is Any Industry Safe?" https://hbswk.hbs.edu/item/amazoned-is-any-industry-safe.
2. Honan, M. (2013). "Remembering the Apple Newton's Prophetic Failure and Lasting Impact," https://www.wired.com/2013/08/remembering-the-apple-newtons-prophetic-failure-and-lasting-ideals.
3. Alter, L. (2014). "The Autoped Was the World's First Scooter." https://www.treehugger.com/bikes/autoped-was-worlds-first-scooter.html.
4. Gerstner, L. (2002). *Who Says Elephants Can't Dance, Leading a Great Enterprise Through Dramatic Change*. New York: Harper Business.
5. Weinberger, M. (2015). "Satya Nadella: 'Customer Love' Is a Better Sign of Success Than Revenue or Profit." *Business Insider*. https://www.businessinsider.com/microsoft-ceo-satya-nadella-on-culture-2015-10.
6. McGrath, R. (2019). *Seeing Around Corners: How to Spot Inflection Points in Business Before They Happen*. New York: Houghton Mifflin Harcourt.
7. CBInsights (2019). "$1B+ Market Map: The World's 390+ Unicorn Companies In One Infographic." https://www.cbinsights.com/research/unicorn-startup-market-map.
8. Schwab, K. (2016). "The Fourth Industrial Revolution: What It Means, How to Respond." https://www.weforum.org/agenda/2016/01/the-fourth-industrial-revolution-what-it-means-and-how-to-respond.
9. Espindola, D. (2019). "Technology Is Exponential but Humans Are Linear." https://www.interceptinghorizons.com/post/technology-is-exponential-but-humans-are-linear Copyright by David sEspindola. Adapted with permission.
10. Intercepting Horizons. https://www.interceptinghorizons.com.
11. O'Reilly III, C. and Tushman, M. (2004). "The Ambidextrous Organization." https://hbr.org/2004/04/the-ambidextrous-organization

Chapter 1 – The New Context for Our Future

12. Bartlett, A. (2013). "Arithmetic, Population and Energy – A Talk by Al Barlett." https://youtu.be/O133ppiVnWY.
13. Quick, L. and Platt, D. (2015). "Disrupted, Strategy for Exponential Change." *Resilient Futures Media*: 17.

14. Scalabrini, A., Ebisch, S.J.H., Huang, Z., and Di Plinio, S. (2019). Spontaneous Brain Activity Predicts Task-Evoked Activity During Animate Versus Inanimate Touch. *Cerebral Cortex* Vol. 29, No. 11: 4628–4645. https://doi.org/10.1093/cercor/bhy340.

15. Wright, M. and Ferguson, W. (2005). *The New Business Normal – The Peril and Promise of the New Global Realities*. Knowledge Management Press.

16. Karakas, F. (2009). "Welcome to World 2.0: The New Digital Ecosystem." *Journal of Business Strategy* Vol. 30, No. 4: 23–30.

17. Qin, P. and Northoff, G. (2011). "How is Our Self Related to Midline Regions and the Default-mode Network?" https://www.sciencedirect.com/science/article/abs/pii/S1053811911005167.

18. McGonical, J. (2017). "Our Puny Human Brains Are Terrible at Thinking About the Future." https://slate.com/technology/2017/04/why-people-are-so-bad-at-thinking-about-the-future.html.

19. Hartley, M. "The Rice and Chessboard Story." *Dr. Mike's Math Games for Kids*. http://www.dr-mikes-math-games-for-kids.com/rice-and-chessboard.html.

20. Gartner. "Gartner Hype Cycle." https://www.gartner.com/en/research/methodologies/gartner-hype-cycle.

21. Grove, A. (1996). *Only the Paranoid Survive: How to Exploit the Crisis Points That Challenge Every Company and Career*. New York: Doubleday.

22. Webb, A. (2016). *The Signals Are Talking: Why Today's Fringe Is Tomorrow's Mainstream*. New York: PublicAffairs.

23. CISCO (2020). Cisco Annual Internet Report (2018–2023) White Paper. https://www.cisco.com/c/en/us/solutions/collateral/executive-perspectives/annual-internet-report/white-paper-c11-741490.html.

24. Gent, E. (2020). "*$100 Genome Sequencing Will Yield a Treasure Trove of Genetic Data*." Singularity Hub. https://singularityhub.com/2020/03/08/100-genome-sequencing-will-yield-a-treasure-trove-of-genetic-data-and-maybe-a-dystopian-nightmare.

25. Rosenberg, M. (2017). "*Marc My Words: The Coming Knowledge Tsunami*." Learning Solutions. https://learningsolutionsmag.com/articles/2468/marc-my-words-the-coming-knowledge-tsunami.

26. RealClear (2019). "Your Mobile Phone vs. Apollo 11's Guidance Computer." https://www.realclearscience.com/articles/2019/07/02/your_mobile_phone_vs_apollo_11s_guidance_computer_111026.html.

27. eTeknix (2017). "Man Drives F1 Car with the Power of His Mind." https://www.eteknix.com/man-drives-f1-car-with-the-power-of-his-mind.

28. Chui, M., Lund S. and Gumbel P. (2018). "*How Will Automation Affect Jobs, Skills and Wages?*" McKinsey Global Institute. https://www.mckinsey.com/featured-insights/future-of-work/how-will-automation-affect-jobs-skills-and-wages.

Chapter 2 – Exponential Platforms: Convergences and Megatrends

29. Diamandis, P. and Kotler, S. (2020). *The Future is Faster Than You Think: How Converging Technologies Are Transforming Business, Industries, and Our Lives*. Simon and Schuster.

30. Foster, R. and Kaplan, S. (2001). *Creative Destruction*. Crown Business.

31. Weintraub, P. (2016) "The New Mind Control." *Aeon*. https://aeon.co/essays/how-the-internet-flips-elections-and-alters-our-thoughts.

32. Kurzweil, R. (2018). "3 Dangerous Ideas from Ray Kurzweil." https://singularityhub. com/2017/11/10/3-dangerous-ideas-from-ray-kurzweil.

33. Redfield, R. (2018). "CDC Director's Media Statement on U.S. Life Expectancy." *CDC Newsroom*. https://www.cdc.gov/media/releases/2018/s1129-US-life-expectancy.html.

34. Netburn, D. (2017). "New Gene Editing Technique May Lead to Treatment of Thousands of Diseases." *LA Times*. https://latimes.com/science/sciencenow/la-sci-sn-dna-gene-editing-20171025-story.html.

35. IamVR Official (2016). "Whale Jumps into a Gym in Mixed Reality (Exciting) by Magic Leap." https://www.youtube.com/watch?v=LM0T6hLH15k.

36. Bailenson, J. (2019). *"The Virtues of Virtual Reality: How Immersive Technology Can Reduce Bias."* NYU Law School. https://www.youtube.com/watch?v=vXxfkkINq8M

37. Diamandis, P. (2020). "20 Megatrends for the Roaring 20s." https://www.diamandis.com/blog/20-metatrends-2020s

38. Peters, A. (2018). *"There Will Soon Be a Whole Community of Ultra-Low-Cost 3-D Printed Homes."* Fast Company. https://www.fastcompany.com/90317441/there-will-soon-be-a-whole-community-made-of-these-ultra-low-cost-3d-printed-homes.

39. Fernholz, T. (2018). "Are There Bubbles in Space." *Quartz*. https://qz.com/1343920/investors-have-pumped-nearly-1-billion-into-aerospace-start-ups-this-year.

40. Hao, K. (2020) "Google Is Using AI to Design Chips That Will Accelerate AI." *MIT Technology Review*. https://www.technologyreview.com/2020/03/27/950258/google-ai-chip-design-reinforcement-learning.

41. Diamandis, P. "The Spatial Web – Part 1," https://www.diamandis.com/blog/the-spatial-web-part-1

42. Kavilanz, P. (2020). "A New Use for McDonald's Used Cooking Oil: 3D Printing." *CNN Business*. https://www.cnn.com/2020/02/19/business/mcdonalds-oil-3d-printing/index.html

43. Espindola, D. (2018). "Artificial Intelligence is Upon Us – Are We Ready?" *Intercepting Horizons*. https://www.interceptinghorizons.com/post/artificial-intelligence-is-upon-us-are-we-ready. Copyright by David Espindola. Adapted with permission.

44. The Next Rembrandt. https://www.nextrembrandt.com.

45. Olewitz, C. (2016). "A Japanese AI Program Just Wrote a Short Novel, and It Almost Won a Literary Prize." *Digital Trends*. https://www.digitaltrends.com/cool-tech/japanese-ai-writes-novel-passes-first-round-nationanl-literary-prize.

46. Kinsey, L. and Azhar, A. (2020). "AI Currents – AI Research You Should Know About." *Exponential View*. https://spark.adobe.com/page/8r3T7EAqvFx7w.

47. Iansiti, M. and Lakhani, K. (2020). *Competing in the Age of AI – Strategy and Leadership When Algorithms and Networks Run the World*. Harvard Business School Publishing.

48. Gartner Newsroom (2019). "Gartner Predicts 90% of Current Enterprise Blockchain Platform Implementations Will Require Replacement by 2021." https://www.gartner.com/en/newsroom/press-releases/2019-07-03-gartner-predicts-90-of-current-enterprise-blockchain.

49. Tuomisto, H. and de Mattos, J.M.T. (2011). "Environmental Impacts of Cultured Meat Production." *Environmental Science and Technology* Vol. 45, No. 14: 6117–6123. https://doi.org/10.1021/es200130u.

50. Strobl, T. "Solving Real-World Problems with Quantum Computing." *BMI*. https://www.businessmodelsinc.com/solving-real-world-problems-with-quantum-computing.

51. Semiconductor Digest (2017). "Number of Connected IoT Devices Will Surge to 125 Billion by 2030." https://sst.semiconductor-digest.com/2017/10/number-of-connected-iot-devices-will-surge-to-125.-billion-by-2030.

52. Kellner, T. (2018). "Fired Up: GE Successfully Tested Its Advanced Turboprop Engine with 3-D Printed Parts." *General Electric.* https://www.ge.com/news/reports/ge-fired-its-3d-printed-advanced-turboprop-engine.

53. Guardian (2015). "Chinese Construction Firm Erects 57-Storey Skyscraper in 19 Days." https://www.theguardian.com/world/2015/apr/30/chinese-construction-firm-erects-57-storey-skyscraper-in-19-days.

54. Claire, S. "Prellis Biologics Reaches Record Speed and Resolution in Viable 3-D Printed Human Tissue." 3-DPrint.com. https://3dprint.com/217267/prellis-biologics-record-speed.

55. Domino's (2016). "Introducing DOM." *Youtube.* https://youtu.be/rb0nxQyv7RU.

56. Mathews, K. (2018). "5 Ways Retail Robots Are Disrupting the Industry." *Robotics Business Review.* https://www.roboticsbusinessreview.com/retail-hospitality/retail-robots-disrupt-industry.

57. Margaritoff, M. (2018). "Drone Deliveries Really Are Coming Soon, Officials Say." *The Drive.* https://www.thedrive.com/tech/19239/drone-deliveries-rally-are-coming-soon-officials-say

58. Wang, B. (2019). "First Commercial Perovskite Solar Late in 2019 and the Road to Moving the Energy Needle." *Next Big Future.* https://www.nextbigfuture.com/2019/02/first-commercial-perovskite-solar-late-in-2019-and-the-road-to-moving-the-energy-needle.html.

Chapter 3 – Animals of the Exponential Kingdom

59. Andreessen, M. (2011). "Why Software is Eating the World." *Wall Street Journal.* https://www.wsj.com/articles/SB10001424053111903480904576512250915629460.

60. Espindola, D. (2018). "Flash Boiled Frogs in the Era of Exponential Change – Are You Feeling the Heat?" *Intercepting Horizons.* https://www.interceptinghorizons.com/post/flash-boiled-frogs-in-the-era-of-exponential-change. Copyright by David Espindola. Adapted with permission.

61. CB Insights. "The Global Unicorn Club." https://www.cbinsights.com/research-unicorn-companies.

62. Aydin, R. (2019). "How 3 Guys Turned Renting Air Mattresses in their Apartment into a $31 Billion Company, Airbnb." *Business Insider.* https://www.businessinsider.com/how-airbnb-was-founded-a-visual-history-2016-2.

63. Iansiti, M. and Lakhani, K. (2020). *Competing in the Age of AI – Strategy and Leadership When Algorithms and Networks Run the World.* Harvard Business Publishing.

64. Byford, S. (2018). "How China's Bytedance Became the World's Most Valuable Startup." *The Verge.* https://www.theverge.com/2018/11/30/18107732/bytedance-valuation-tiktok-china-startup.

65. Kharpal, A. (2019). "TikTok Owner ByteDance is a $75 Billion Chinese Tech Giant – Here's What You Need to Know About It." *CNBC.* https://www.cnbc.com/2019./05/30/tiktok-owner-bytedance-what-to-know-about-the-chinese-tech-giant.html.

66. Singh, M. (2019). "Bytedance, Tiktok's Parent Company, Plans to Launch A Free Music Streaming App." *TechCrunch.* https://techcrunch.com/2019/05/21/tiktok-bytedance-music-app.

67. Liu, N. "China's ByteDance Plans to Develop its Own Smartphone." *Financial Times.* https://www.ft.com/content/ea4a2be4-7dca-11e9-81d2-f785092ab560.

68. South China Morning Post (2018). "Meet the 35-year-old Chinese Software Engineer Behind ByteDance, the World's Most Valuable Start-up." *Bloomberg.* https://www.scmp.com/tech/start-ups/article/2166424/meet-35-year-old-chinese-software-engineer-behind-bytedance-worlds.

69. Number8. (2017). "What is a 'Gazelle' Company?" https://number8.com/gazelle-company.

70. Blystone, D. (2019). "The Story of Uber". *Investopedia.* https://www.investopedia.com/articles/personal-finance/111015/story-uber.asp.

71. Diamandis, P. and Kotler S. (2020). *The Future is Faster Than You Think – How Converging Technologies Are Transforming Business, Industries, and Our Lives.* Simon & Schuster.

72. Castillo, M. (2017). "Reed Hastings' story about the founding of Netflix has changed several times." *CNBC.* https://www.cnbc.com/2017/05/23/netflix-ceo-reed-hastings-on-how-the-company-was-born.html.

73. McFadden, C. (2019). "The Fascinating History of Netflix." *Interesting Engineering.* https://interestingengineering.com/the-fascinating-history-of-netflix.

74. Wingfield, N. and Stelter, B. (2011). "How Netflix Lost 800,000 Members and Good Will." *New York Times.* https://www.nytimes.com/2011/10/25/technology/netflix-lost-800000-members-with-price-rise-and-split-plan.html.

75. Gaudet, C. "6 Strategies Netflix Can Teach Us for Dominating Our Market." *Predictable Profits.* https://predictableprofits.com/6-strategies-netflix-can-teach-us-dominating-market.

76. Mohammed, S. (2019). "IBM's Turnaround Under Lou Gerstner – Business & Management Lessons, Case Study." *Medium.* https://medium.com/@shahmm/ibms-turnaround-under-lou-gerstner-case-study-business-management-lessons-a0dcce04612d.

77. Graham, P. (2007). "Microsoft is Dead." http://www.paulgraham.com/microsoft.html

78. Nadella, S. (2017). *Hit Refresh – The Quest to Rediscover Microsoft's Soul and Imagine a Better Future for Everyone.* Harper Business.

79. McFadden, C. (2018). "Almost Everything You Need to Know About Google's History." *Interesting Engineering.* https://interestingengineering.com/almost-everything-you-need-to-know-about-googles-history

80. Simpson, C. (2013). "Google Wants to Cheat Death." *The Atlantic.* https://www.theatlantic.com/technology/archive/2013/09/google-wants-cheat-death/310943.

81. Kim, E. (2018). "Amazon Echo owners Spend More on Amazon than Prime Members, report says." *CNBC.* https://www.cnbc.com/2018/01/03/amazon-echo-owners-spend-more-on-amazon-than-prime-members.html

82. Lomas, N. and Crook, J. (2017). "Amazon has Acquired 3D Body Model Startup, Body Labs, for $50M-$70M." *Techcrunch.* https://techcrunch.com/2017/10/03/amazon-has-acquired-3d-body-model-startup-body-labs-for-50m-70m.

83. Shead, S. (2017). "Amazon Now has 45,000 Robots in its Warehouses." *Business Insider.* https://www.businessinsider.com/amazons-robot-army-has-grown-by-50-2017-1

84. Boyle, A. (2019). "Amazon to Offer Broadband Access from Orbit with 3,236 Satellite 'ProjectKuiper' Constellation." *Geekwire.* https://www.geekwire.com/2019/amazon-project-kuiper-broadband-satellite

Chapter 4 – Introducing SPX: Strategic Planning for the Exponential Era

85. Wilson, N. (2019). On the Road to Convergence Research. *Bioscience* Vol. 69 No. 8: 587–593.

86. Beck, K. et al.(2001) "Manifesto for Agile Software Development." https://agilemanifesto.org

87. Lker, J. (2004). *The Toyota Way*. McGraw-Hill Education.
88. Ries, E. (2011). *The Lean Startup*. Crown Business.
89. Burnett, B. and Evans, D. (2018). *Designing Your Life*. Alfred Knopf.
90. Ideo U "What is Design Thinking?" https://www.ideou.com/blogs/inspiration/what-is-design-thinking.
91. Wikipedia. OODA Loop. https://en.wikipedia.org/wiki/OODA_loop.

Chapter 5 – Detecting Early Signals

92. NASA Jet Propulsion Laboratory. California Institute of Technology. Catching a Whisper from Space. https://www.jpl.nasa.gov/edu/teach/activity/catching-a-whisper-from-space.
93. The Free Beginners Guide. 3D Printing Industry. https://3dprintingindustry.com/3d-printing-basics-free-beginners-guide#02-history.
94. Steven Blank (2016). "Why Tim Cook Is Steve Ballmer and Why He Still Has His Job at Apple," https://steveblank.com/2016/10/24/why-tim-cook-is-steve-ballmer-and-why-he-still-has-his-job-at-apple.
95. Bossidy, L. and Charan (2002). *Execution: The Discipline of Getting Things Done*. Crown Business.
96. Blank, S. (2007). *The Four Steps to the Epiphany: Successful Strategies for Products That Win*. Third Edition. Quad/Graphics.
97. McGrath R., McManus, R and Fujitsu (2020). "The Swift Rise and Surprising Importance of Digital Ecosystems." https://www.fujitsu.com/downloads/GLOBAL/vision/2019/download-center/FTSV2019_wp4_EN_1.pdf
98. Webb, A. (2020). "The 11 Sources of Disruption Every Company Must Monitor." *MIT Sloan Management Review*. https://sloanreview.mit.edu/article/the-11-sources-of-disruption-every-company-must-monitor
99. Weinberger, M. (2015). "Satya Nadella: 'Customer Love' Is a Better Sign of Success Than Revenue or Profit." *Business Insider*. https://www.businessinsider.in/satya-nadella-customer-love-is-a-better-sign-of-success-than-revenue-or-profit/articleshow/49266047.cms.
100. Marr, B. (2018) "How Much Data Do We Create Every Day? The Mind-Blowing Stats Everyone Should Read." *Forbes*. https://www.forbes.com/sites/bernardmarr/2018/05/21/how-much-data-do-we-create-every-day-the-mind-blowing-stats-everyone-should-read/#57bf26960ba9

Chapter 6 – Learning from Experiments

101. Epstein, D. (2019). *Range: Why Generalists Triumph in a Specialized World*. New York: Riverhead Books.
102. Griffin, A., Price, R. L. and Vojak, B. (2012). *Serial Innovators: How Individuals Create and Deliver Breakthrough Innovations in Mature Firms*. Redwood City: Stanford Business Books.
103. Netflix Technology Blog. (2016). "It's All A/Bout Testing: The Netflix Experimentation Platform." https://netflixtechblog.com/its-all-a-bout-testing-the-netflix-experimentation-platform-4e1ca458c15.
104. Ferriss, T. (2007). *The 4-Hour Workweek*. New York: Crown.
105. Ries, E. (2011). *The Lean Startup*. New York: Crown Business.

106. (2007). "Netflix to offer online movie viewing." CNN Money." https://money.cnn.com/2007/01/16/technology/netflix.
107. McGrath, R. (2019). *Seeing Around Corners*. New York: Houghton Mifflin Harcourt.

Chapter 7 – Capabilities – The Essential Fuel to Ride the Exponential Curve

108. Bossidy, L. and Charan, R. (2002). *Execution: The Disciplines of Getting Things Done*. New York: Crown Business.
109. Rumsfeld, D. (2013). *Rumsfeld's Rules*. New York: HarperCollins.
110. Iansiti, M. and Lakhani, K. (2020). *Competing in the Age of AI: Strategy and Leadership When Algorithms and Networks Run the World*. Boston: HBR Press.
111. Ramadan, A. et al. (2016). *Play Bigger – How Pirates, Dreamers, and Innovators Create and Dominate Markets*. New York:HarperCollins.
112. Faeste, L. and Hemerling, J. (2016). *Transformation: Delivering and Sustaining Breakthrough Performance*. Boston: The Boston Consulting Group.

Chapter 8 – Feedback-Based Strategic Decisions

113. Rumsfeld, D. (2013). *Rumsfeld's Rules*. New York: HarperCollins.
114. Ries, E. (2011). *The Lean Startup*. New York: Crown Business.
115. Christian, B. and Griffiths, T. (2016). *Algorithms to Live By*. New York: Picador.
116. Iansiti, M. and Lakhani, K. (2020). *Competing in the Age of AI: Strategy and Leadership When Algorithms and Networks Run the World*. Boston: HBR Press.
117. Glaveski, S. (2020). "12 Business Lessons from Seth Godin." https://medium.com/swlh/12-business-lessons-from-seth-godin-acb0e71e1adc.
118. Taleb, N. (2007). *The Black Swan: The Impact of the Highly Improbable*. New York: Random House.
119. Bossidy, L. and Charan, R. (2002). *Execution: The Discipline of Getting Things Done*. New York: Crown Business.

Chapter 9 – Leading a Culture of Change

120. Nadella, S., Shaw, G. and Nichols, J.T. (2017). *Hit Refresh: The Quest to Rediscover Microsoft's Soul and Imagine a Better Future for Everyone*. New York: HarperCollins.
121. Dweck, C. (2007). *Mindset: The New Psychology of Success*. New York: Ballantine Books.
122. Christensen, C. (2016). *The Innovator's Dilemma: When New Technologies Cause Great Firms to Fail*. Boston: Harvard Business Review Press.
123. Christian, B. and Griffiths, T. (2016). *Algorithms to Live By: The Computer Science of Human Decisions*. New York: Picador.
124. Sims, P. (2013). *Little Bets: How Breakthrough Ideas Emerge from Small Discoveries*. New York: Simon & Schuster.
125. McCracken, H. (2017). "Satya Nadella Rewrites Microsoft's Code." Fast Company. https://www.fastcompany.com/40457458/satya-nadella-rewrites-microsofts-code

126. Levitt, S. (2016). "Heads or Tails: The Impact of a Coin Toss on Major Life Decisions and Subsequent Happiness." NBER Working Paper No. 22487.
127. Godin, S. (2007). *The Dip: A Little Book That Teaches You When to Quit (and When to Stick).* New York: Portfolio.
128. Duckworth, A. et al. (2016). "Don't Grade Schools on Grit." New York Times.
129. Konnikova, M. (2017). *The Confidence Game: Why We Fall for It. . .Every Time.* New York: Penguin Books.
130. Intercepting Horizons. https://www.interceptinghorizons.com.
131. Wright, M. and Ferguson, W. (2005). *The New Business Normal – The Peril and Promise of New Global Realities.* New York: Knowledge Management Press
132. Netflix's jobs website. https://jobs.netflix.com/culture.

Chapter 10 – The Exponential Human – Social and Ethical Challenges in a Chaotic World

133. McLeod, V. (2020). "COVID-19: A History of Coronavirus." Lab Manager. https://www.labmanager.com/lab-health-and-safety/covid-19-a-history-of-coronavirus-22021.
134. Susskind, D. (2020). *A World Without Work – Technology, Automation, and How We Should Respond.* New York: Metropolitan Books.
135. Diana, F. (2020). "A World Without Work." Reimagining the Future. https://frankdiana.net/2020/02/10/a-world-without-work.
136. Passy, J. and Keshner, A. (2020). "Harvard and Other Major Universities Still Charging Full Tuition as Classes Go Online Amid Coronavirus Outbreak." MarketWatch. https://www.marketwatch.com/story/a-lot-of-colleges-simply-cant-afford-to-give-refunds-major-universities-holding-online-classes-due-to-coronavirus-are-still-charging-full-tuition-2020-03-13.
137. McGrath, R. (2019). *Seeing Around Corners – How to Spot Inflection Points in Business Before They Happen.* New York: Houghton Mifflin Harcourt.
138. Diamandis, P. and Kotler, S. (2020). *The Future is Faster Than You Think – How Converging Technologies Are Transforming Business, Industries, and Our Lives.* New York: Simon and Schuster.
139. Copyright by Michael W. Wright. Reprinted with permission.
140. Lewis, T. (2019). "AI Can Read Your Emotions. Should It?" The Guardian. https://www.theguardian.com/technology/2019/aug/17/emotion-ai-artificial-intelligence-mood-realeyes-amazon-facebook-emotient
141. Mahmood, M. et al. (2019). "Fully Portable and Wireless Universal Brain–Machine Interfaces Enabled by Flexible Scalp Electronics and Deep Learning Algorithm." *Nature Machine Intelligence* Vol. 1: 412–422. https://doi.org/10.1038/s42256-019-0091-7
142. Chandler, S. (2019). "Brain Computer Interfaces and Mind Control Move One Step Closer to Becoming Reality." *Forbes.* https://www.forbes.com/sites/simonchandler/2019/09/24/brain-computer-interfaces-and-mind-control-move-one-step-closer-to-becoming-reality/#1a88a70f32fb
143. Wheatley, M. (2003). "Willing to be Disturbed." *Kaos Pilot A-Z* by Uffe Ubaek. Aarhus, Denmark: KaosCommunication.
144. Copyright by Michael W. Wright. Reprinted with permission.
145. Copyright by Michael W. Wright. Reprinted with permission.
146. Copyright by Michael W. Wright. Reprinted with permission.
147. Copyright by Michael W. Wright. Reprinted with permission.
148. Copyright by Michael W. Wright. Reprinted with permission.

INDEX

The Exponential Era: Strategies to Stay Ahead of the Curve in an Era of Chaotic Changes and Disruptive Forces, First Edition. David Espindola and Michael W. Wright.
© 2021 by The Institute of Electrical and Electronics Engineers, Inc.
Published 2021 by John Wiley & Sons, Inc.